Biomechanics of the Wrist Joint

Kai-Nan An
Richard A. Berger
William P. Cooney III
Editors

Biomechanics of the Wrist Joint

With 88 Illustrations

Springer-Verlag
New York Berlin Heidelberg London Paris
Tokyo Hong Kong Barcelona Budapest

Kai-Nan An
Richard A. Berger
William P. Cooney III
Mayo Clinic
Rochester, MN 55905
USA

Library of Congress Cataloging-in-Publication Division
Biomechanics of the wrist joint / [editors], Kai-Nan An, Richard A. Berger, William P. Cooney.
p. cm
Includes bibliographical references.
ISBN-13:978-1-4612-7833-7
1. Wrist--Mechanical properties. 2. Biomechanics. I. An, Kai-Nan. II. Berger, Richard A., 1954- . III. Cooney, William Patrick, 1943-
[DNLM: 1. Biomechanics. 2. Wrist Joint-physiology. WE 830 B6155]
QP334.B56 1991
612.7'5--dc20
DNLM/DLC
for Library of Congress 91-5055
Printed on acid-free paper.

Softcover reprint of the hardcover 1st edition 1991

Camera-ready copy provided by the editors.

9 8 7 6 5 4 3 2 1

ISBN-13:978-1-4612-7833-7 e-ISBN-13:978-1-4612-3208-7
DOI: 10.1007/978-1-4612-3208-7

We dedicate this book to our parents and families who support us and to all the past, present and future investigators interested in the wrist.

Foreword

Clinical interest in the wrist joint has accelerated markedly in the last two decades. Clinical diagnosis based on a greater understanding of wrist anatomy, biomechanics and increasingly sophisticated imaging techniques has markedly enhanced our ability to treat disorders of this joint. As our clinical acumen becomes better, we increasingly need more accurate understanding of the basic mechanisms by which the wrist is able to carry out its function. This book represents a compendium of work done by a number of authors in the basic sciences and their presentations at a recent workshop on biomechanics. This work, while at the forefront of current research in this area, is but an indicator of the type of information that is increasingly required to progress in this field. The authors have made some sound contributions and this book should be of considerable interest and help to those individuals who are contributing to progress in this field. It will be of even greater importance if it helps to stimulate the reader to become involved in further research into the intricacies of the wrist and help us to solve its numerous problems. I hope the reader will enjoy reading these chapters as much as I did in listening to them at the time of their presentations.

Ronald L. Linscheid, M.D.
President 1989-1990
American Society for Surgery of the Hand
Mayo Clinic
Rochester, Minnesota

Preface

Work related injury has become a major factor in current world economics. In the United States, disabilities related to injury of the hand and wrist rank second only to repetitive stress trauma as a cause of lost work days. In spite of this, there are still many controversies and difficulties in the clinical diagnosis and treatment of the problem of the wrist joint. In order to attempt to resolve these difficulties, many investigations of the functional anatomy and biomechanics of the wrist joint have been recently performed and published.

The wrist joint is not like any other joint in the body. It is not a simple hinge joint or ball-socket joint and does not have an idea mechanical equivalent. Each of the carpal bones has its own unique center or axis of rotation and that wrist motion which we can observe and measure is, in fact, a result of the combination of the small carpal bones moving on each other. This interactions of the carpal bones is conceptually analogous to a Rubic's cube, in which motion in one segment directly affects the position of the another segment. By the same token, force transmission through the wrist joint occurs in a very complex manner.

Studies of wrist joint motion and force transmission have posed very challenging engineering problems. Numerous sophisticated experimental and analytic methods have been developed or adopted in attempts to increase our understanding of the biomechanics of the wrist joint. This has been especially true over the past fifteen years, where we have witnessed a tremendous surge of biomechanic research performed on the wrist joint. Included in these studies have been the experimental measurements and analytic calculations of the individual carpal bone motion in-vitro and in-vivo under normal and pathological conditions. These methods have also been used successfully to compare various treatment modalities and surgical procedures. Similarly, numerous experimental and analytic methods have also been instrumented for the determination of the force transmission and pressure distribution on the articular surfaces under normal and abnormal conditions.

In order to critically review various engineering methods, synthesize the available findings and their clinical implications and applications, and to define the direction for the future research, a workshop format gathering of various investigators currently engaged in wrist biomechanics studies was recently held. The results of this unique gathering have been compiled to form the basis of this book. We believe the documentation of such information will not only be beneficial for individuals interested in wrist biomechanics and clinicians treating wrist joint problems, but will also of interest to those involved in joint mechanics research in general.

biomechanics and clinicians treating wrist joint problems, but will also of interest to those involved in joint mechanics research in general.

This book has been written to present various aspects of wrist biomechanics in nine chapters. General anatomy pertaining to those structures felt to be important to our understanding of biomechanics are briefly summarized in Chapter 1. In the next three chapters, the kinematics of the wrist joint involved in various activities and the movement of individual carpal bones are presented. Some of the basic concepts related to kinematic analysis are also included in Chapter 2. Force transmission through the wrist joint as a unit as well as individual carpal bones based on the analytic and experimental methods, are included in Chapters 5, 6 and 7. Finally, to complete the topic of static and dynamic balance of the wrist joint, the material and mechanical properties of the ligaments and the biomechanical function of muscles are presented in Chapters 8 and 9 respectively.

K-N. An, Ph.D.
R.A. Berger, M.D., Ph.D.
W.P. Cooney III, M.D.
Mayo Clinic
Rochester, Minnesota

Contents

Contributors

K-N. An, Ph.D. Orthopedic Biomechanics Laboratory, Mayo Medical School/Mayo Clinic, Rochester, Minnesota, 55905, U.S.A.

R.A. Berger, M.D., Ph.D. Departments of Orthopedic Surgery and Anatomy, Mayo Medical School/Mayo Clinic, Rochester, Minnesota, 55905, U.S.A.

E.Y-S. Chao, Ph.D. Orthopedic Biomechanics Laboratory, Mayo Medical School/Mayo Clinic, Rochester, Minnesota, 55905, U.S.A.

W.P. Cooney, III, M.D. Department of Orthopedic Surgery, Mayo Medical School/Mayo Clinic, Rochester, Minnesota, 55905, U.S.A.

M. Garcia-Elias, M.D. Department of Orthopedic Surgery, Hospital General de Catalunya, Barcelona, Spain

E. Horii, M.D. Orthopedic Department, Branch Hospital of Nagoya University, University of Nagoya, Nagoya City, Japan

V.R. Masear, M.D. Department of Orthopedic Surgery, University of Alabama, Birmingham, Alabama, 35233, U.S.A.

M.D. Nowak, Sc.D. Orthopaedic Research, Department of Orthopaedics, University of Connecticut Health Center, Farmington, Connecticut, 06032, U.S.A.

A.K. Palmer, M.D. Department of Orthopedic Surgery, State University of New York, Health Science Center, Syracuse, New York, 13210, U.S.A.

R.M. Patterson, M.Eng. Division of Orthopaedic Surgery, University of Texas Medical Branch at Galveston, Galveston, Texas, 77550, U.S.A.

J. Ryu, M.D. Department of Orthopaedic Surgery, Texas Tech University Health Sciences Center, El Paso, Texas, 79905, U.S.A.

S.F. Viegas, M.D. Division of Orthopaedic Surgery, University of Texas Medical Branch at Galveston, Galveston, Texas, 77550, U.S.A.

F.W. Werner, M.M.E. Department of Orthopedic Surgery, State University of New York, Health Science Center, Syracuse, New York, 13210, U.S.A.

Contributors

Acknowledgments for Permissions

The editors wish to thank the publishers and authors listed below for their copyright permission and endorsement to use their previously published figures in this book. Their valuable help in this matter has made the publication possible. The figures and/or pictures listed below are reprinted with permission from the following:

Chapter 1:

Figure 1 reprinted with permission from Journal of Hand Surgery from Figure 1 in Volume 15A, pp. 847-854, 1990; Mosby-Year Book, Inc.

Chapter 2:

Figure 1 reprinted with permission from Annals of Biomedical Engineering, from Figure 1 in Volume 12, pp. 585-597, 1984; Pergamon Press Ltd.

Figure 2 reprinted with permission from W.B. Saunders Company from Figure 6-12 in "The Shoulder," 1990, by C.A. Rockwood and F.A. Matsen.

Figure 3 reprinted with permission from Annals of Biomedical Engineering from Figure 2 in Volume 12, pp. 585-597, 1984; Pergamon Press Ltd.

Figure 4 reprinted with permission from Annals of Biomedical Engineering from Figure 3 in Volume 12, pp. 585-597, 1984; Pergamon Press PLC.

Figure 5 reprinted with permission from Hand Clinics from Figure 1 in Volume 6, pp. 393-403, 1990; W.B. Saunders Company.

Chapter 3:

Figures 1, 2, 3, 4, 5, 6, 7, and 8 reprinted with permission from Journal of Hand Surgery from Figures 1, 2, 3, 4, 5, 6, 7, and 8 in Volume 16A, pp. 409-419, 1991; Mosby-Year Book, Inc.

Chapter 4:

Figure 1 reprinted with permission from Journal of Orthopaedic Research from Figures 1 and 2 in Volume 7, pp. 590-598, 1989; Raven Press, Ltd.

Figure 2 reprinted with permission from Journal of Orthopaedic Research from Figure 3 in Volume 7, pp. 590-598, 1989; Raven Press, Ltd.

Figures 3, 4, 5, and 6 reprinted with permission from Journal of Hand Surgery from Figures 5, 3, 7, and 8 in Volume 14A, pp. 791-799, 1989; Mosby-Year Book, Inc.

Chapter 5:

Figure 2 reprinted with permission from Journal of Hand Surgery from Figure 1 in Volume 15A, pp. 393-400, 1990; Mosby-Year Book, Inc.

Figure 3 reprinted with permission from Journal of Orthopaedic Research from Figure 2 in Volume 7, pp. 738-743, 1989; Raven Press, Ltd.

Figure 5 reprinted with permission from Journal of Hand Surgery from Figure 1 in Volume 12A, pp. 196-202, 1987; Mosby-Year Book, Inc.

Figure 8 reprinted with permission from Hand Clinics from Figure 13 in Volume 3, pp. 31-40, 1987; W.B. Saunders Company.

Figure 9 reprinted with permission from Clinical Orthopaedics and Related Research from Figure 9 in Volume 187, pp. 26-35, 1984; J.B. Lippincott Company.

Chapter 6:

Figures 1 and 2 reprinted with permission from Journal of Hand Surgery from Figures 6 and 8 in Volume 12A, pp. 971-978, 1987; Mosby-Year Book, Inc.

Figures 3,4 and 5 reprinted with permission from Journal of Hand Surgery from Figures 5,4 and 8 in Volume 14A, pp. 458-465, 1989; Mosby-Year Book, Inc.

Figure 7 reprinted with permission from Journal of Hand Surgery from Figures 9 in Volume 12A, pp. 978-985, 1987; Mosby-Year Book, Inc.

Figure 8 reprinted with permission from Journal of Hand Surgery from Figures 1,2 and 3 in Volume 15A, pp. 268-278, 1990; Mosby-Year Book, Inc.

Chapter 9:

Figures 1 and 2 reprinted with permission from Journal of Biomechanics from Figures 1 and 5 in Volume 22, pp. 943-948, 1989; Pergamon Press PLC.

Chapter 1

General Anatomy of the Wrist

R.A. Berger and M. Garcia-Elias

Introduction

In recent years, there has been increasing interest in the anatomic and functional complexities of the wrist. With the assumption that to understand the mechanics of a joint a thorough knowledge of its anatomy must be sought, a number of investigations have been performed to clarify the spatial relationships between the different soft tissue components of the wrist and their associated joints (Berger, et al., 1982; Berger, et al., 1984; Berger and Blair, 1984; Berger and Landsmeer, 1990; Berger, et al., 1991; Burgess 1990; Cihâc 1972; Cooney, et al., 1989; Drewniany, et al., 1985; Garcia-Elias and Domenech-Mateu, 1987; Garcia-Elias, et al., 1989; Jessurun, et al., 1987; Kauer 1974; Kauer 1975; Kauer 1980; Landsmeer 1976; Lewis, et al., 1970; Linscheid 1986; Mayfield, et al., 1976; Mayfield, et al., 1979; Roger, et al., 1985; Skahen, et al., 1990; Taleisnik 1976; Weber 1984).

Different techniques have been utilized: gross and microscope dissections (Berger and Blair, 1984; Berger, et al., 1982; Berger and Landsmeer, 1990; Drewniany, et al., 1985; Lewis, et al., 1970; Mayfield, et al., 1976; Taleisnik 1976; Testut 1928), specimen cross-sectional studies (Garcia-Elias, et al., 1989; Weber 1984), histological analysis of embryos (Cihâc 1972), fetuses (Berger and Landsmeer, 1990; Berger, et al., 1991; Garcia-Elias and Domenech-Mateu, 1987; Kauer 1975; Landsmeer 1976) and cadaver specimens (Berger and Blair, 1984; Berger and Landsmeer, 1990; Cooney, et al., 1989; Kauer 1980), arthrotomography (Berger, et al., 1984), arthroscopy

(Bottle, et al., 1989; North and Thomas, 1988), computed tomography (Jessurun, et al., 1987; Roger, et al., 1985) and magnetic resonance imaging of living subjects (Skahen, et al., 1990), etc. The goal of these techniques has been not only to better describe the different structures involved in wrist function, but also to quantify their static and dynamic spatial relationships (location, direction, stiffness, strength, etc.) (Berger, et al., 1982; Drewniany, et al., 1985; Garcia-Elias, et al., 1989; Linscheid 1986; Weber 1984).

This chapter will examine the pertinent skeletal, joint and ligamentous anatomy of the wrist, updating classical textbook descriptions with data obtained from the most recent investigations in this field. The distal radioulnar joint, though certainly involved in wrist function, is now being considered part of the so-called "forearm articulation" (Hagert 1987) and therefore, will not be covered in this chapter.

Skeletal Anatomy

Although traditionally described as a single joint, the wrist is a composite articulation with overall motion resulting from the summation of interactions of the individual carpal bones amongst themselves as well as distally with the bases of the metacarpals and proximally with the distal articulating surface of the radius and the ulna/triangular fibrocartilage complex. Generally speaking, the eight carpal bones can be conveniently divided into two anatomic rows, a proximal and distal carpal row. Beginning radially, the proximal carpal row is composed of the scaphoid, lunate, triquetrum and pisiform, while the distal row consists of the trapezium, trapezoid, capitate and hamate. Some authors prefer to exclude the pisiform from consideration as a true carpal bone, as it is a sesamoid within the tendon of flexor carpi ulnaris. The radiocarpal joint is formed by the articulation of the proximal carpal row and the distal articulating surface of the radius and triangular fibrocartilage complex. The mid-carpal joint is that articulation found between the proximal and distal rows. Individual joint clefts are found between the mutual articulating surfaces of the bones within each carpal row. It should be noted that an additional schematic division of the carpus has been proposed by Navarro and revised by Taleisnik which divides the carpus into radial, central and ulnar columns (Taleisnik 1976). This scheme is based more on theoretical considerations of function rather than anatomic associations.

The distal articular surface of the radius is concave and tilted in two planes with an average of 11° of palmar tilt in the sagittal plane and an average of 22° of ulnar inclination in the coronal plane. In a wrist with neutral ulnar variance, the transition of the distal articular surface of the radius to the TFCC is smooth and nearly indistinguishable. A consistent surface marking on the distal articular surface of the radius is the interfacet prominence, which is a fibrocartilaginous sagittal ridge, progressively more prominent volarly, which separates the distal articulating surface of the radius into lunate and scaphoid fossae. The scaphoid fossa is triangular and is larger than the more quadrangular lunate fossa (Figure 1).

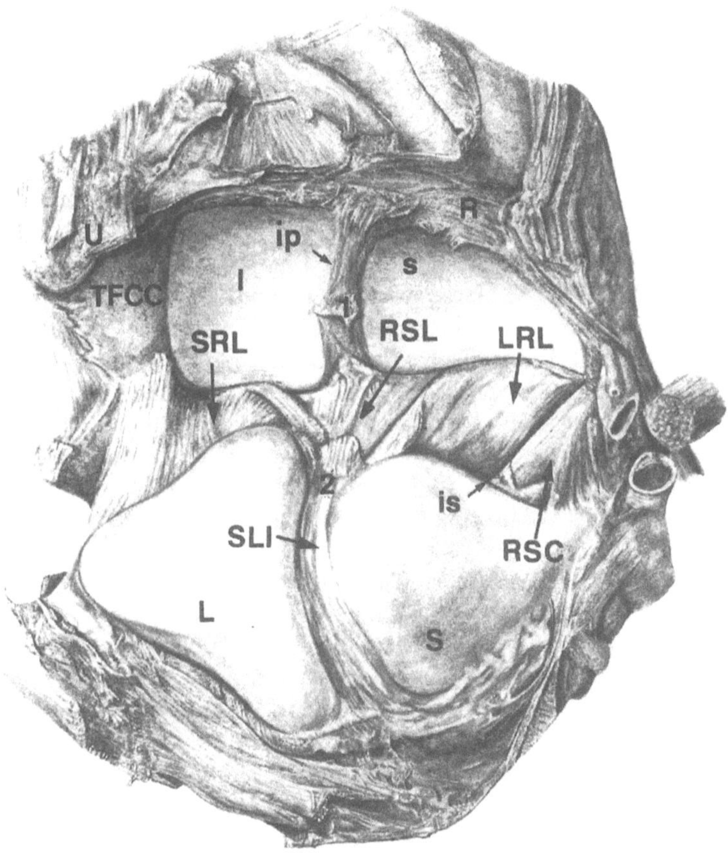

FIGURE 1: Drawing of the radiocarpal joint from a dorsal and distal perspective with the dorsal capsule incised and the proximal row palmar flexed. R = radius, U = ulna, S = scaphoid, L = lunate, s = scaphoid fossa, l = lunate fossa, IP = interfacet prominence, RSC = radioscaphocapitate ligament, LRL = long radiolunate ligament, RSL = radioscapholunate ligament, SRL = short radiolunate ligament, is = interligamentous sulcus. Reprinted with permission: Berger RA and Landsmeer JMF, J Hand Surg 15A:847-854, 1990.

The triangular fibrocartilage complex (TFCC) is composed of a fibrocartilaginous disc, interposed between the head of the ulna and the proximal carpal row, which is supported anteriorly and posteriorly by the palmar and dorsal radioulnar ligaments, respectively (Palmer and Werner, 1981). Palmarly and ulnarly, it is continuous with ulnocarpal ligaments, and dorsally contributes to the tendon sheath of extensor carpi ulnaris. A small aperture is consistently found in the TFCC just proximal to the junction of the medial fibers of the TFCC where they attach to the medial border of the triquetrum. This aperture is called the prestyloid recess and variably communicates with the tip of the ulnar styloid process. The second aperture is variably found (30-60 percent of normal wrists) just proximal to the palmar aspect of the triquetrum, which represents a communication between the radiocarpal and pisotriquetral joints. Both the prestyloid recess and the communication to the pisotriquetral articulation are lined by synovial villi.

The radiocarpal joint is normally isolated from communication with the distal radioulnar joint and the mid-carpal joint by virtue of competent interosseous membranes between the bones of the proximal carpal row and a competent TFCC. Other than communication with the pisotriquetral joint, communication of fluid between the radiocarpal joint and either the distal radioulnar joint or mid-carpal joint is considered abnormal. There is evidence, however, that such communications may represent a progressive degenerative change related to age rather than a significant destabilizing event associated with trauma, although the latter may certainly occur.

Each carpal bone is distinctly unique in shape and thus contributes uniquely to the mechanism of the wrist. However, as a generalization, each carpal bone may be represented schematically as a cube with each face of the cube dedicated to articular contact or capsular attachment. The marginal carpal bones (trapezium, scaphoid, hamate and triquetrum), in general, have three faces of the cube dedicated to capsular attachment, thus leaving three faces for articular contact. In contrast, the central bones (capitate, trapezoid and lunate) will have only the dorsal and palmar surfaces available for capsular attachment with the remaining four sides covered in large part by articular cartilage. There are no predictable tendinous attachments to the carpal bones with the single exception of the pisiform. There may be one or more aberrant slips of the abductor pollicis longus tendon which inserts into the trapezium. Some authors consider the pisotriquetral ligament an anatomic extension of the tendon of flexor carpi ulnaris, thus the attachment to the hamulus of the hamate may be considered tendinous insertion. There are few areas on the surface of the carpal bones which are not covered by articular cartilage or direct capsular insertions, such as the neck of the capitate and the palmar surface of the proximal pole of the scaphoid. These areas are lined by extensions of the synovial layer of the joint capsule.

In the radiocarpal joint, the proximal articulating surface of the scaphoid has a higher radius of curvature than that of the lunate. This is reflected in differences in curvature of the distal articular surface of the radius. The proximal articular surface of the triquetrum is relatively flat, but in large part does not articulate with another bone. Rather, it rests against the TFCC. The mid-carpal joint is a combination of three different types of articulation. Laterally, the convex distal surface of the scaphoid articulates with the concavity formed by the trapezium, trapezoid and the lateral aspect of the capitate. The central part of the mid-carpal joint is concave proximally, formed by the distal surfaces of the scaphoid and lunate and convex distally, formed by the head of the capitate and variably the proximal pole of the hamate. Medially, the triquetrohamate articulation is helicoid or screw-shape in configuration. Consideration of the geometry of the carpometacarpal joints is beyond the scope of this monograph.

Ligamentous Anatomy

Most of the ligaments of the wrist are considered true intracapsular ligaments (Berger and Landsmeer, 1990). By definition, this implies that the ligamentous tissue, composed of longitudinally oriented fascicles of collagen, is found between the fibrous and synovial layers of the joint capsule. With the exception of the pisotriquetral ligament and the flexor and extensor retinacula, there are no extracapulsar ligaments of the wrist. As will be discussed later, however, there are ligaments which do not fit the definition of intracapsular ligaments in that they are completely surrounded by a synovial lining, and thus, are intra-articular ligaments.

By Taleisnik's definition, the carpal ligaments may be divided into extrinsic and intrinsic groups based upon their location of origin (Taleisnik 1976). Extrinsic ligaments have an attachment either proximal or distal to the carpal bones in addition to an attachment on the carpal bones, while intrinsic ligaments attach entirely within the confines of the carpus. Numerous descriptions of the carpal ligaments have been proposed in the past. The description offered here represents the most current understanding of ligamentous anatomy.

Extrinsic Ligaments

The palmar wrist joint capsule completely covers the carpal bones and joint spaces. After dissecting away the synovial lining of the carpal tunnel, the remaining fibers appear to originate from the radial and ulnar borders of the distal forearm and form two V-shaped ligamentous bands, one proximal connecting the forearm to the proximal row and one distal linking the forearm to the distal carpal row. More recently, however, a number of anatomic studies have demonstrated further subdivisions of the structures. The individual ligaments can be more clearly defined when viewed from within the wrist through either a dorsal capsulotomy with a hyperflexed wrist or arthroscopically.

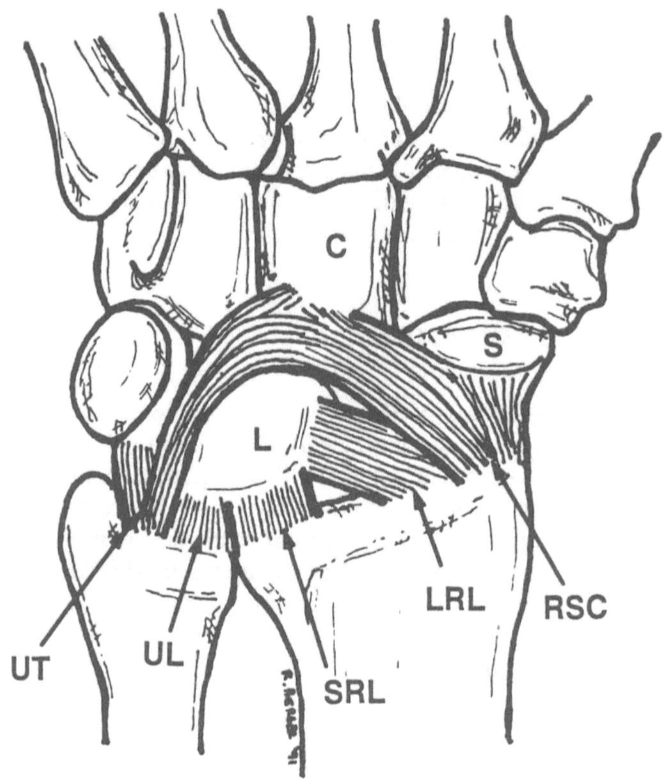

FIGURE 2: Schematic of the carpal region from a palmar perspective with the extrinsic palmar carpal ligaments illustrated. S = scaphoid, C = capitate, L = lunate, RSC = radioscaphocapitate ligament, LRL = long radiolunate ligament, SRL = short radiolunate ligament, UL = ulnolunate ligament, UT = ulnotriquetral ligament. Note how the medial most fibers of the RSC ligament arch around the distal aspect of the palmar horn of the lunate to interdigitate with fibers from the UL/UT complex to form the arcuate ligament which supports the head of the capitate. Only a small percentage of fibers from this complex insert into the body of the capitate.

Beginning radially, four distinct extrinsic palmar radiocarpal ligaments have been defined. Originating from the radial styloid process and the radial most palmar lip of the radius is the *radioscaphocapitate (RSC) ligament* (Figures 1 and 2). It courses distally as a single ligament, but can be divided artificially into three components. The radial most component inserts onto the lateral aspect of the waist of the scaphoid to behave anatomically as a radial collateral ligament, however this functional label has been challenged and has stirred significant controversy. Medially, contiguous fibers of the radioscaphocapitate ligament insert hemicircumferentially about the proximal surface of the distal pole of the scaphoid. The most medial fibers of the radioscaphocapitate ligament course deep to the proximal pole of the scaphoid, obliquely toward mid-carpal articulation, where they interdigitate with fibers emanating from the triangular fibrocartilage complex and the triquetrum to form a supporting "sling" for the head of the capitate, sometimes referred to as the *arcuate ligament*. Only a small percentage of fibers from the RSC ligament insert into the body of the capitate. The region of interdigitation of fibers from the RSC ligament and those from the TFCC and triquetrum is found just distal to the palmar horn of the lunate, being separated from the palmar horn of the lunate by the Space of Poirier. This ligamentous support for the anterior aspect of the head of the capitate may act as a mechanical sling preventing mid-carpal palmar subluxation of the capitate during normal wrist extension.

Just ulnar to the origin of the RSC ligament, with a small amount of palmar overlap, the *long radiolunate (LRL) ligament* connects the radius to the lunate (Figures 1 and 2). This ligament in the past has been called the radiolunotriquetral ligament, however new evidence suggests that there is an insufficient amount of ligament coursing continuously over the palmar horn of the lunate to continue to the triquetrum to justify referring to this ligament by this term. The LRL ligament lies palmar to the proximal pole of the scaphoid and supports it, theoretically, much in the same way as the arcuate ligament supports the head of the capitate. The LRL ligament inserts in the radial half of the palmar horn of the lunate. The RSC and LRL ligaments are separated throughout their course by a deep division called the interligamentous sulcus (Figures 1 and 3). This forms an important landmark for use at arthrotomy or during arthroscopy and is continuous with the Space of Poirier more distally.

Just ulnar to the origin of the LRL ligament, the *radioscapholunate (RSL) ligament* enters the radiocarpal joint space through a defect in the palmar radiocarpal joint capsule (Figures 1 and 3). This ligament has been referred to in the past as the radioscaphoid, radiolunate, radioscapholunate ligament, as well as the ligament of Testut (Testut 1928). Recent anatomic studies utilizing both adult and fetal histologic sections have shown that this ligament is not a true connective tissue ligament, but is in actuality a neurovascular vincula supplied by branches from the palmar carpal branch of the radial artery and the anterior interosseous artery and nerve (Berger and Blair, 1984; Berger, et al., 1991). It is covered by a thick synovial lining readily appreciated using an arthroscope. The RSL ligament is continuous with the membranous proximal portion of the scapholunate ligament and attached to the interfacet prominence. Fetal studies have suggested that this represents a vestige of a septum which divides the radiocarpal joint into radioscaphoid and radiolunate clefts (Berger, et al., 1991; Lewis, et al., 1970). The mechanical contributions of the radioscapholunate ligament are hypothesized to be minimal, a concept which is supported by the ligament load failure test of Mayfield, et al, (1979) who demonstrated that this ligament resists only modest tensile loads.

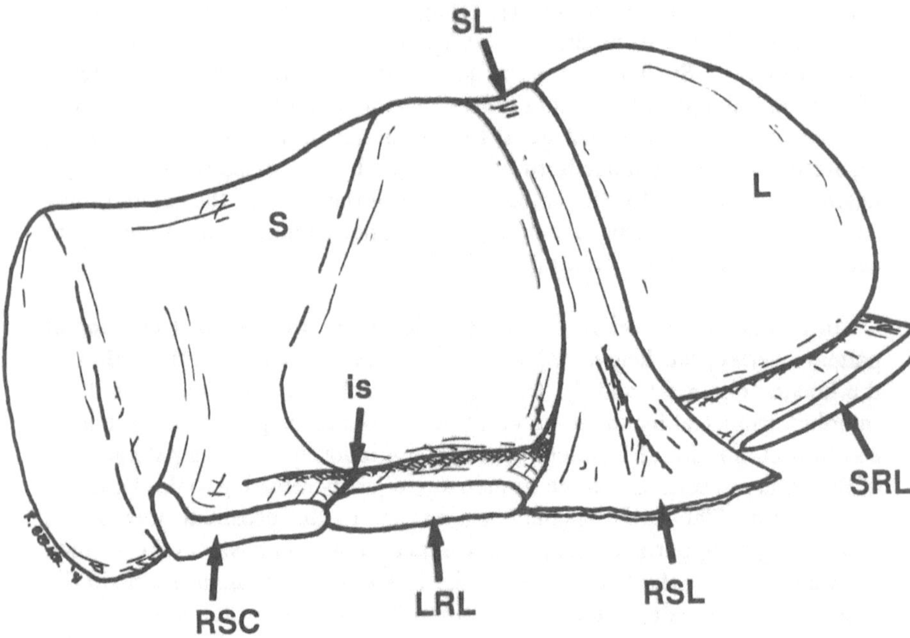

FIGURE 3: Drawing of the scapholunate complex from a proximal and radial perspective. S = scaphoid, L = lunate, RSC = radioscaphocapitate ligament, LRL = long radiolunate ligament, RSL = radioscapholunate ligament, SRL = short radiolunate ligament, SL = scapholunate interosseous ligament, is = interligamentous sulcus. Note how the radioscapholunate ligament is shown penetrating through the palmar capsule between LRL and SRL and how it forms in part the proximal membranous aspect of SL.

Arising from the palmar margin of the lunate fossa, and coursing distally as a flat, yet stout, ligament to insert into the proximal aspect of the palmar horn of the lunate is the *short radiolunate (SRL) ligament* (Figures 1 and 2). The short radiolunate ligament is consistently separated from the long radiolunate ligament by the penetration of the radioscapholunate ligament through the palmar carpal capsule. Ulnarly, the short radiolunate ligament blends imperceptibly with fibers emanating from the palmar aspect of the TFCC, which also attach to the lunate. The short radiolunate ligament appears to be a principal stabilizer of the lunate and is often the only viable soft tissue attachment holding the lunate to the distal radius in anterior lunate dislocations.

The *ulnolunate (UL) ligament* (Figure 2), in direct continuity with the SRL ligament, has a principal attachment to the proximal aspect of the palmar horn of the lunate. More ulnarly, however, a modest percentage of fibers course distally, anterior to the palmar portion of the lunotriquetral ligament, where they begin to arc radially to merge with fibers from the RSC ligament palmar to the head of the capitate to form the *arcuate ligament*. Just ulnar to the ulnolunate ligament, the *ulnotriquetral (UT) ligament* inserts principally on the medial surface of the triquetrum, anatomically appearing as an ulna collateral ligament (Figure 2). The fibers continue distally to insert onto the medial surface of the hamate, interdigitating with fibers of the triquetrohamate ligament. In 60 to 70 percent of normal wrists, the communication between the radiocarpal and pisotriquetral joints identifies the anatomic division between the ulnolunate and ulnotriquetral components of the ulnocarpal ligament complex.

The only extrinsic ligament on the dorsum of the carpus is the *dorsal radiocarpal (DRC) ligament*, which, as on the palmar side, is a true intracapsular ligament (Figure 4). It has been described as being formed by two components: a superficial radiotriquetral band and a deep radiolunotriquetral ligament. The superficial component is wide and thin, originating from the dorsal margin of the distal radius, centered just distal to Lister's tubercle. It courses obliquely distal and ulnarly and inserts on the dorsal rim of the triquetrum. There are some more vertically oriented fibers arising from the dorsal distal border of the radial notch which course distally to insert into a triangular rough facet located proximal to the dorsal tubercle of the triquetrum. Inseparable in most specimens from the superficial band, the deep component of the dorsal radiocarpal ligament arises from the medial third of the distal and dorsal border of the radius and courses obliquely to insert into the distal part of the lunotriquetral articulation, intermingling with fibers of the lunotriquetral ligament. Aside from the deep virtue fascicles of the dorsal radiocarpal ligament, there is not a well differentiated and consistent dorsal radiolunate ligament.

Intrinsic Ligaments

Interdigitating with the triquetral insertion of the DRC ligament, an intrinsic ligament called the *dorsal intercarpal (DIC) ligament* originates to course distally and radially to insert on the dorsal surface of the waist and the distal pole of the scaphoid and to a lesser degree on the dorsal surface of the trapezoid (Figure 4). It passes just distal to the dorsal horn of the lunate and forms the floor of the fourth and fifth extensor compartments as they cross the wrist region. The ligament is thickest proximally, and has some interconnections with the dorsal lunotriquetral ligament and the dorsal scapholunate ligament. This structure probably has an important role in the transverse stabilization of the proximal carpal row.

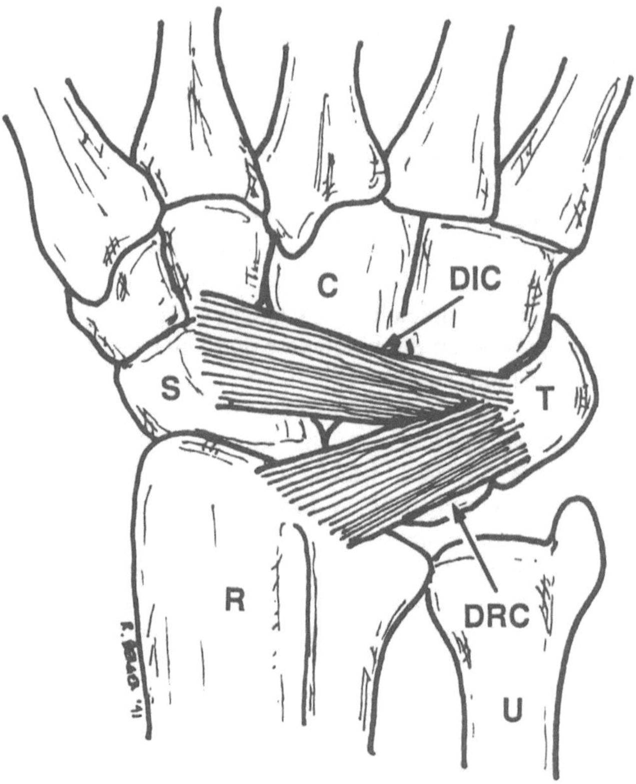

FIGURE 4: Drawing of the carpal region from a dorsal perspective. R = radius, U = ulna, S = scaphoid, C = capitate, T = triquetrum, DRC = dorsal radiocarpal ligament, DIC = dorsal intercarpal ligament.

There are very few ligaments which actually cross the mid-carpal joint, which may be in part responsible for the relatively great mobility between the proximal and distal carpal rows. Stability, however, remains a key feature to normal mechanics of this joint, and this is in part dependent on the following intrinsic ligaments. On the palmar surface of the carpus, beginning radially, the first intrinsic mid-carpal ligament is the palmar *scaphotrapeziotrapezoidal (STT) ligament* (Figure 5). These fibers originate from the palmar surface of the distal pole of the scaphoid, particularly its distal aspect, and diverge to insert onto the proximal surface the palmar tubercle of the trapezium and the proximal palmar surface of the trapezoid. Just ulnar to the origin of the STT ligament, the *scaphocapitate (SC) ligament* originates from the distal pole of the scaphoid (Figure 5). The proximal margin of the SC ligament is contiguous with the distal margin of the RSC ligament. This is a substantial ligament which courses obliquely, distally and ulnarly to insert onto the radial half of the palmar surface of the capitate. The STT and SC ligaments have been shown in a recent investigation to play an important role as major distal stabilizers of the scaphoid.

In the ulnar region of the carpus, the mid-carpal intrinsic ligament complex is formed by the *triquetrohamate (TH) ligament* and *triquetrocapitate (TC) ligament* (Figure 5). The TH ligament originates from the distal margin of the palmar surface of the triquetrum, just radial to the pisotriquetral joint capsule. It courses distally to insert onto the palmar surface of the body of the hamate. There is some interdigitation with the ulnotriquetral ligament at the base of the hook of the hamate. The TC ligament is formed by fibers diverging radially from the TH ligament to insert on the ulnar half of the palmar surface of the body of the capitate. Because of their apparent convergence over the mid-line of the capitate, the mid-carpal intrinsic ligaments originating from the scaphoid and triquetrum have received the generic name of the arcuate ligament of the wrist.

The intrinsic ligaments within the proximal and distal carpal rows are unique, discrete, and probably play a substantial role in maintaining the mechanical integrity of each row respectively. As a generalization, these ligaments are composed of dorsal and palmar regions and interconnect no more than two adjacent carpal bones.

Within the proximal carpal row, two intrinsic interosseous ligaments are found: the *scapholunate (SL)* and *lunotriquetral (LT) ligaments* (Figures 5 and 6). Both ligaments are composed of true ligaments dorsally and palmarly as defined by histologic studies, which show collinear fascicles of collagen, and a proximal membranous region, connecting the dorsal and palmar ligaments, which is composed of fibrocartilage (Figure 7). This is also referred to as the interosseous membrane. The palmar region of the SL ligament is longer than the dorsal region and has a more oblique orientation, perhaps allowing more rotation between the two bones. The membranous region of the SL ligament is continuous with the radioscapholunate ligament. The lunotriquetral ligament is quite thick dorsally and palmarly with essentially transversely oriented fascicles. When intact, the membranous regions of the SL and LT ligaments isolate the mid-carpal joint from the radiocarpal joint.

Unlike the proximal row interosseous ligaments, the distal carpal row interosseous ligaments do not form a system which isolates the mid-carpal joint from the carpometacarpal joint. There are no membranous regions of these ligament. Individual ligaments, although difficult to separate anatomically, are found on the dorsal and palmar surfaces of the joint spaces,

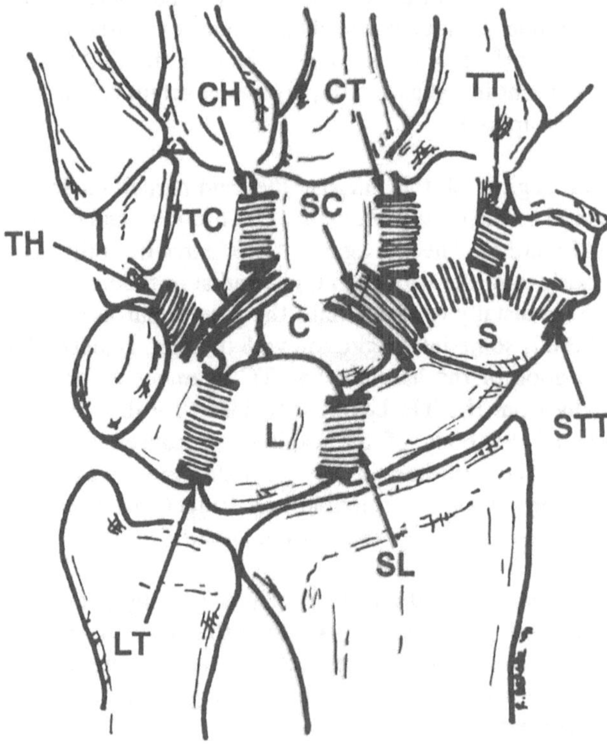

FIGURE 5: Drawing of the carpal region from a palmar perspective showing the palmar intrinsic ligaments. S = scaphoid, L = lunate, C = capitate, SL = scapholunate ligament, LT = lunotriquetral ligament, STT = scaphotrapeziotrapezoidal ligament, SC = scaphocapitate ligament, TC = triquetrocapitate ligament, TH = triquetrohamate ligament, TT = trapeziotrapezoid ligament, CT = capitotrapezoid ligament, and CH = capitohamate ligament. SL, LT, TT, CT, and CH represent the palmar aspects of the interosseous ligaments connecting bones within the proximal and distal carpal rows. STT, SC, TC, and TH represent the intrinsic ligaments that span the mid-carpal joint on the palmar side.

and are called the *trapeziotrapezoid (TT)*, *capitotrapezoid (CT)* and *capitohamate (CH) ligaments* (Figures 5 and 6). The CT and CH ligaments are present only on the distal 50 percent of the surface of the capitate. There are two "deep" intrinsic ligaments in the distal carpal row which are not appreciated until the respective joints are opened. The *deep capitohamate (DCH) ligament* is found in a nearly square recess in the contiguous joint surfaces of the capitate and hamate in the palmar and distal corner of the articular surfaces (Figure 8). This is a substantial ligament, averaging approximately 25 mm in cross-sectional area. Similarly, the *deep capitotrapezoid (DCT) ligament* is situated in the mutual articular surfaces of the trapezoid and capitate, angling obliquely in a mutual recess (Figure 8). No specific investigations regarding the mechanical contributions of these deep ligaments has been published. The size of the ligaments suggest a significant stabilizing contribution to the distal carpal row and histologically they are both noted to have a significant content of nerve tissues suggesting the possibility of proprioceptor function as well.

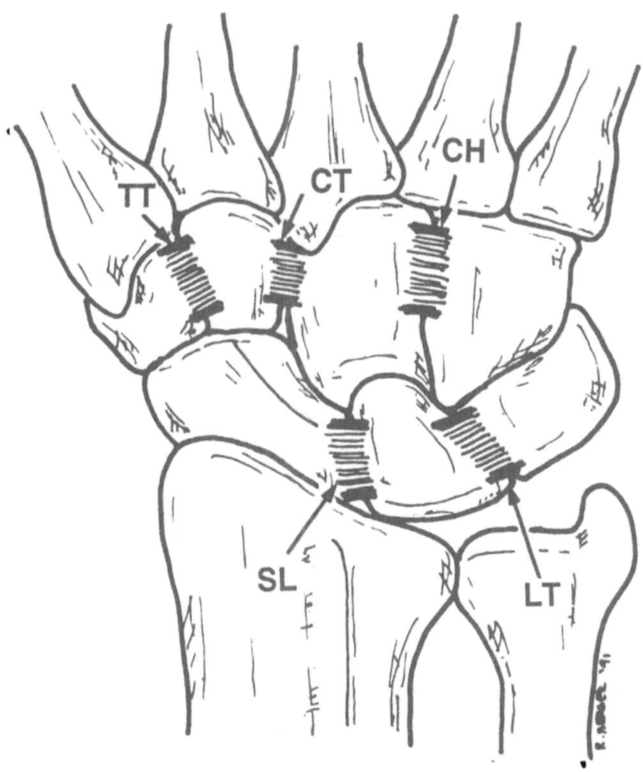

FIGURE 6: Drawing of the carpal region from a dorsal perspective showing the dorsal intrinsic ligaments. SL = scapholunate ligament, LT = lunotriquetral ligament, TT = trapeziotrapezoid ligament, CT = capitotrapezoid ligament, and CH = capitohamate ligament. The only dorsal ligament consistently found which crosses the mid-carpal joint is the DIC ligament, not shown in this drawing (see Figure 4).

Retinacular System of the Wrist

Only one forearm muscle, the flexor carpi ulnaris (FCU) has its major distal insertion on the carpus. It attaches to the pisiform which, in its turn, is connected to the hook of the hamate via the pisihamate ligament, and to the fifth metacarpal, by means of the pisometacarpal ligament. The flexor carpi radialis (FCR) tendon inserts on the palmar aspect of the base of the second metacarpal. The wrist extensors [extensor carpi radialis longus (ECRL), extensor carpi radialis brevis (ECRB) and extensor carpi ulnaris (ECU)] insert on the dorsal surface of the base of the second, third and fifth metacarpal bones, respectively. Other tendons crossing the wrist joint are: the abductor pollicis longus (APL) and the extensor pollicis brevis (EPB), laterally; the extensor pollicis longus (EPL), the extensor digitorum communis (EDC), the extensor indicis proprius (EIP), and the extensor digiti quanti (EDQ), dorsally; and the flexor digitorum sublimis (FDS) and profundae (FDP) and the flexor pollicis longus (FPL), palmarly.

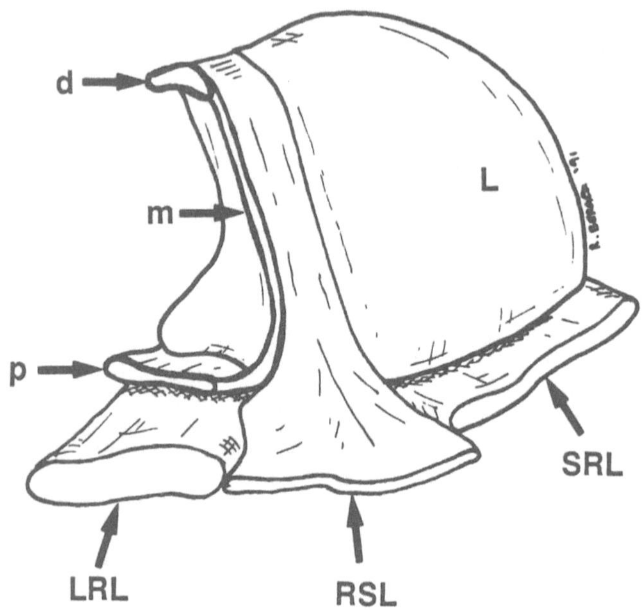

FIGURE 7: Drawing of the scapholunate complex from a proximal and radial perspective with the scaphoid excised. L = lunate, LRL = long radiolunate ligament, RSL = radioscapholunate ligament, SRL = short radiolunate ligament. The scapholunate ligament, which merges proximally and palmarly with the RSL ligament can be divided into three regions as follows: d = the thick dorsal ligamentous portion, p = the thin palmar ligamentous portion, and m = the mid-fibrocartilaginous region. Note that the palmar scapholunate ligament is distinctly separate from the long radiolunate ligament.

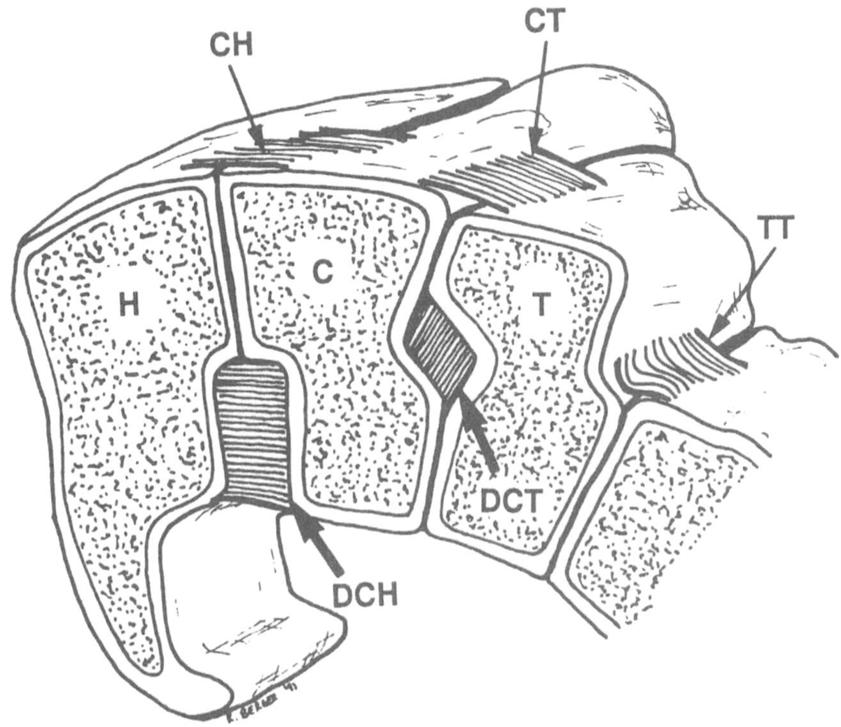

FIGURE 8: Drawing of the distal carpal row from a distal and radial perspective, with a simulated osteotomy through the bones just proximal to the carpometacarpal joint. H = hamate, C = capitate, T = trapezoid, CH = dorsal capitohamate ligament, CT = dorsal capitotrapezoid ligament, TT = dorsal trapeziotrapezoid ligament. DCH is the deep capitohamate ligament, found within the capitohamate joint, with transversely oriented fibers. The palmar aspect of DCH is continuous with the palmar capitohamate ligament. DCT is the deep capitotrapezoid ligament, also found within the joint space, with obliquely oriented fibers as shown. The DCT ligament is completely independent from any dorsal or palmar intrinsic ligaments.

In order to maintain a normal relationship between the tendons crossing the wrist joint and its capsular structures there is a complex retinacular system. It derives from the pars profunda of the antebrachial fascia, and can be divided in two parts: flexor and extensor retinaculum.

Flexor Retinaculum

The deep layer of the forearm fascia begins to thicken in the distal forearm where it receives arcuate tendinous expansions emerging from the FCU, FCR, and palmaris longus tendons. At the level of the proximal carpal row, this fibrous layer thickens considerably, forming the palmar carpal ligament which attaches laterally on the scaphoid tuberosity and medially on the pisiform. More distally, at the level of the distal carpal row, the fibers of the flexor retinaculum form the tranverse carpal ligament becoming more transverse and thicker than proximal fibers. They connect the hook of the hamate to the palmar ridge of the trapezium, forming the palmar closure to the so-called carpal tunnel, a fibrous compartment containing all flexor tendons (except the FCU and FCR) and the median nerve (Figure 9). The floor of the carpal tunnel is the fibrous layer covering the palmar radiocarpal ligaments. The FCR tendon is not contained in the carpal tunnel with the other flexor tendons, but is found within a separate tunnel delineated by a fibrous sagittal septum, that arises from the palmar aspect of the trapezoid and joins the flexor retinaculum near to its lateral insertion of the trapezium.

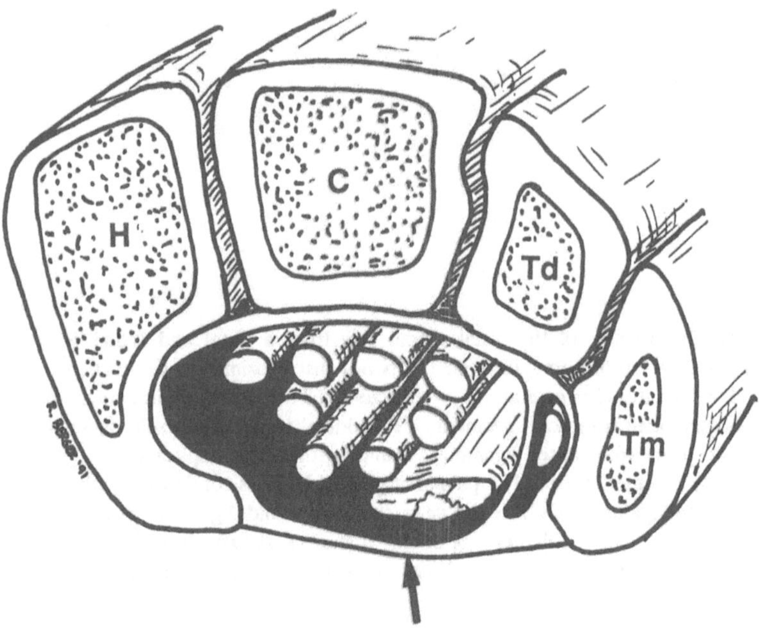

FIGURE 9: Drawing of the distal carpal row from a distal and radial perspective in a cross-section format through the middle of the carpal bones. H = hamate, C = capitate, Td = trapezoid, Tm = trapezium, F.R. = flexor retinaculum. The flexor retinaculum at this level spans the carpal tunnel between the trapezium and hook of the hamate, and is continuous with the periosteum on the nonarticular surfaces of the adjacent carpal bones as well as the fibrous strata of the capsular carpal ligaments. The contents of the carpal tunnel are shown as the flexor tendons and median nerve, with the flexor carpi radialis tendon traveling in its own compartment.

Extensor Retinaculum

The dorsal retinaculum of the wrist is also derived from the deep forearm fascia. It consists of two layers: the supratendinous and the infratendinous layers. Originating from the deep aspect of the supratendinous layer are six longitudinal vertical fibrous septa, inserting into the radius, that create six compartments for the different extensor tendons crossing the wrist (Figure 10).

The first compartment, on the lateral border of the wrist, contains the APL and EPB tendons. Quite often, a septum separates this compartment into two or more channels. The second compartment contains the ECRL and ECRB tendons. The third compartment, which lies ulnar to Lister's tubercle, contains the EPL tendon. The floor of the fourth and fifth compartments, which contain the extensor digitorum communis, extensor indicis proprius, and the extensor digiti quanti, respectively, is the infratendinous retinaculum, a sheet of parallel fibers, narrower and shorter than those of the supratendinous retinaculum, which blend with the dorsal wrist capsule and the corresponding sagittal septa. The sixth compartment, for the ECU tendon, is an independent fibrous tunnel formed exclusively by the infratendinous retinaculum, separated from the supratendinous layer by loose connective tissue.

The fibers of the central part of the supratendinous layer, the thickest part of the extensor retinaculum, insert laterally into the radius, forming the lateral septum for the first compartment. These fascicles run somewhat obliquely over the underlying extensor tendons, attach to the ulnar side of the triquetrum before inserting into the pisiform where they blend with expansions of the FCU tendon and the origins of the abductor digiti quanti.

Suggested Techniques for Anatomic Investigation

Although the discipline of anatomy involves levels of morphology ranging from gross anatomy to ultrastructure, the application of anatomic principles to the musculoskeletal system at a level that is commensurate with biomechanics is limited. Numerous anatomic descriptions of the wrist have surfaced, particularly in the last three decades. It is incumbent upon the anatomist, however, to continually review the anatomic descriptions from the great European anatomists, particularly from the 18th and 19th centuries. A careful review of these classic descriptions reveals a surprising degree of detail which unfortunately has been obscured in recent renditions.

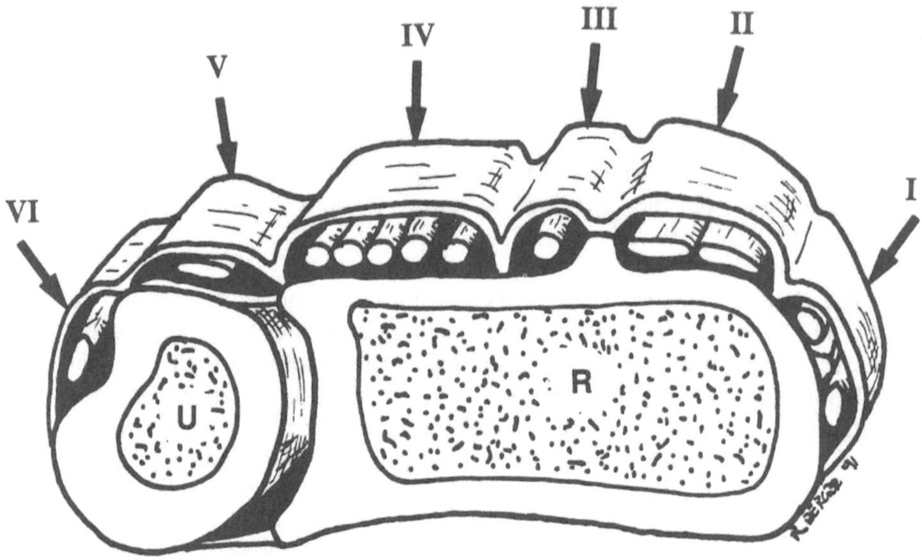

FIGURE 10: Drawing of the distal radius and ulna as a cross-section from a distal and radial perspective illustrating its extensor retinaculum system. I-VI illustrate the six extensor compartments. I = first extensor compartment containing the tendons of abductor pollicis longus and extensor pollicis brevis, II = second extensor compartment containing the tendons of extensor carpi radialis longus and brevis, III = third extensor compartment containing the tendon of extensor pollicis longus, IV = fourth extensor compartment containing the tendons of extensor digitorum communis and extensor indicis proprius, V = fifth extensor compartment containing the tendon of extensor digiti minimi, and VI = sixth extensor compartment containing the tendon of extensor carpi ulnaris.

No doubt the greatest obstacle in setting out to provide a description of a structure in gross anatomic terms lies in the inherent destructiveness of the dissection process itself. If anatomy is to be considered useful, the structure being described must be related to the surrounding structures. This in itself poses a paradox, in that those surrounding structures are generally destroyed in order to gain access to the structure which is to be described. Therefore, the anatomist must be very clear about the region to be described and make all attempts to approach this region in a fashion which will allow an accurate description of the relationship of the objective structure to its surrounding. This can be accomplished in several ways, each with its own set of limitations and advantages. The first is to approach the dissection through a series of layered dissections, carried out in a way that each preceding layer dissected can be returned to its proper anatomic position and orientation as deeper layers are exposed. This system, however, has an inherent error source in that the definition of layers is dependent entirely upon the dissector's discretion. This method of dissection is also dependent upon the direction of approach to the objective structure, and should utilize descriptions based upon approaches from several directions. An example of this would be a description of the triangular fibrocartilage complex based upon approaches from the dorsal, ulnar, palmar, distal and proximal approaches. This can be facilitated by utilizing the second method of anatomic approach which involves the piecemeal excision of structures which are not felt to be necessary for the objective anatomic description. An example of this is a description of the palmar radiocarpal ligaments from a dorsal perspective after piecemeal removal of the carpal bones (Berger and Landsmeer, 1990). It is important, however, to maintain an awareness of the relationship of the excised tissue to the remaining tissue for orientation purposes. Perhaps the best method of obtaining anatomic descriptions at a gross level of structures in the musculoskeletal system, particularly the wrist, employs the concept of serial sections (Berger, et al., 1984; Palmer and Werner, 1981). Although preserved specimens will suffice, it is felt that fresh specimens are far superior. Sections of sufficient quality can be obtained from a thoroughly frozen specimen using a band saw with an adequately sharp blade turning at a fast speed. Some investigators have embedded the frozen specimen in a cube of inert material such as latex prior to sectioning. This step is not absolutely necessary, as it is possible to obtain quality 1-2 mm thick sections in any orientation with an unembedded specimen. In order to fully appreciate the three-dimensional anatomy of the object structure, it is advisable to obtain sections in the transverse, parasagittal and paracoronal planes. This technique is superior in that it preserves all planes of tissue which allows the anatomist to make a complete description of the object structure as well as relevant contiguous structures without having disturbed their natural relationships. This technique is time-consuming, but the results are well worth the effort. Due to the thickness of the sections, however, the investigator is advised to be certain to make observations of the object structures on both sides of the sections created. It is possible to take photographs of these sections and digitize the outlines of specific structures for the purpose of three-dimensional graphic reconstruction. It is also possible to freeze several specimens in different positions, obtain similar sections, and observe differences in the configurations and orientations of the various tissues within the section. This technique is very useful for generating a quantitative description of the geometry of articular surfaces.

Accurate descriptions of the interrelationship of soft tissues may not be possible with gross anatomic techniques alone. In these situations, it is advisable to incorporate the use of light microscopy. For example, an investigation designed to determine the orientation of collagen bundles within a ligament or to study in detail the region of interfacing of one or more ligaments would best be carried out using light microscopic images obtained from serial sections of the tissues (Berger and Blair, 1984). One problem with this historically has been the maintenance of orientation of the soft tissue relative to the bones. A method has been worked out, however, which eliminates this problem by using bone-ligament-bone complexes (Berger, et al., 1982). After en-bloc excision of the bone-ligament-bone complex of interest from a cadaver specimen, the bones are trimmed and cancellous bone is removed using a dental drill bit. The specimens are decalcified in either 5 percent formic acid or similar decalcification agent, fixed by routine techniques, paraffin mounted, and sectioned in 8 to 10 micron sections. Previous investigations have revealed that this technique preserves the orientation of the ligament relative to the bone and is sufficiently gentle to the soft tissues allowing preservation of the adequate morphologic detail. The sections can then be stained with a wide variety of stains, most commonly using hematoxylin and eosin. It is recommended that sections be obtained in planes which are similar to the cardinal planes of gross anatomy (transverse, coronal, and sagittal) in order to relate the findings to gross anatomic descriptions. It is not absolutely necessary to evaluate each serial section, such that in area of high interest, every tenth section can be viewed and in areas of intermediate interest, every 100th or so section can be viewed. It is useful to utilize a microscope with photographic capabilities. As with the gross serial sections, the histologic sections can be digitized for three-dimensional graphic reconstruction (Tinkelenberg 1979). Utilization of this technique, although time-consuming and somewhat laborious, provides a tremendous amount of information in a way which preserves the three-dimensional relationships of the structures of interest relative to surrounding tissues.

Although the material may be difficult to obtain, the usefulness of fetal extremities which have been prepared and sectioned for light microscopy has been demonstrated in a number of recent publications (Berger and Landsmeer, 1990; Berger, et al., 1991; Garcia-Elias and Domenech-Mateu, 1987; Kauer 1975; Landsmeer 1976). The size of the extremity makes overall scanning of the structures much easier than in adult specimens and it is often useful to derive an understanding of changes which occur in the developing extremity through various fetal stages. A word of caution, however, should be issued when making generalizations regarding the relationship of structures in the fetal extremity relative to those same structures in an adult. Proportions, orientation, and possible further tissue differentiation makes such generalizations hazardous.

The presentation of results of anatomic investigations is as important as the method utilized to derive those results and the accuracy of the description. It is through the presentation of results that others will attempt to learn what the anatomist has discovered. Although photographic documentation is considered to be the most irrefutable method of documentation of results, the limitations of photographic methods may diminish the impact of the description. Problems encountered include limited depth of field, limited resolution of fine structures in photographing a large field, and deterioration of detail caused by confusing and unnecessary surrounding tissues left in the image. Therefore, it is very useful to incorporate adjunct illustrations with the photographic recordings. These illustrations require a sense of trust on the part of the observer for the accuracy of the illustrator. Illustrations are very useful in highlighting small objects in a large field of view, three-dimensional structures, superimposed structures, and anatomic descriptions, to name just a few. The close alliance of the anatomic investigator and a skillful and creative medical artist is a relationship worth searching for.

Significant advances in imaging are occurring at a staggering pace. Our understanding of the mechanics of joint systems is also advancing at a rapid rate. It is incumbent upon the investigator to continually review and update as necessary the degree of detail of anatomic descriptions. Although certainly accurate for their level of detail at the time, the anatomic descriptions of the 18th and 19th centuries will soon become insufficient as our ability to understand the subtle mechanics of the wrist joint and our ability to clinically image structures in the wrist joint continues to advance.

References

Berger RA, Kauer JMG, Landsmeer JMR: The radioscapholunate ligament: a gross anatomic and histologic study of fetal and adult wrists. J Hand Surg 1991;16:350-355.

Berger RA, Landsmeer JMF: The palmar radiocarpal ligaments: A study of adult and fetal human wrist joints. J Hand Surg 1990;15A:847-854.

Berger RA, Blair WF: The radioscapholunate ligament: A gross and histologic description. Anat Rec 1984;210:393-405.

Berger RA, Blair WF, El-Khoury GY: Arthrotomography of the wrist: The palmar radiocarpal ligaments. Clin Orthop Rel Res 1984;186:224-229.

Berger RA, Blair WF, Crowninshield RD, Flatt AE: The scapholunate ligament. J Hand Surg 1982;7:87-91.

Berger RA, LaVelle S, Blair, WF, Maynard JA: A technique for the decalcification of large bone-ligament complexes. J Histotechnology 1982;5(4):175-177.

Botte MJ, Cooney WP, Linscheid RL: Arthroscopy of the wrist: Anatomy and technique. J Hand Surg 1989;14A:313.

Burgess RC: Anatomic variations of the mid-carpal joint. J Hand Surg 1990;15A:129-131.

Cihák R: Ontogenesis of the skeleton and intrinsic muscles of the human hand and foot. Ergebnisse der Anatomie und Entwiglungschichte 1972;46:1-189.

Cooney WP, Garcia-Elias M, Dobyns JH, Linscheid RL: Anatomy and mechanics of carpal instability. Surg Rounds for Orthop 1989;3:15-25.
Drewniany JJ, Palmer AK, Flatt AE: The scaphotrapezial ligament complex: An anatomic and biomechanical study. J Hand Surg 1985;10A:492-498.
Garcia-Elias M, An KN, Cooney WP, Linscheid RL, Chao EYS: Stability of the transverse carpal arch: An experimental study. J Hand Surg 1989;14A:277-281.
Garcia-Elias M, Domenech-Mateu JM: The articular disc of the wrist. Limits and relations. Acta Anta 1987;128:51-54.
Hagert C: The distal radioulnar joint. Hand Clinics 1987;3:41-50.
Jessurun W, Hillen B, Zonneveld F, et al.: Anatomical relations in the carpal tunnel: A computed tomographic study. J Hand Surg 1987;12B:64-67.
Kauer JMG: Functional anatomy of the wrist. Clin Orthop 1980;149:9-20.
Kauer JMG: The articular disc of the hand. Acta Anat 1975;93:590-611.
Kauer JMG: The interdependence of carpal articulation chains. Acta Anat 1974;88:481-501.
Landsmeer JMF: Atlas of anatomy of the hand. New York, Churchill Livingstone, 1976.
Lewis OJ, Hamshere RJ, Bucknill TM: The Anatomy of the wrist joint. J Anatomy 1970;106:539-552.
Linscheid RL: Kinematic consideration of the wrist. Clin Orthop 1986;202:27-39.
Mayfield JK, Williams WJ, Erdman AG, et al.: Biomechanical properties of human carpal ligaments. Orthop Trans 1979;3:143-144.
Mayfield JK, Johnson RP, Kilcoyne RF: The ligaments of the human wrist and their functional significance. Anat Rec 1976;186:417-428.
Mizuseki T, Ikuta Y: The dorsal carpal ligaments: Their anatomy and function. J Hand Surg 1989;14B:91-98.
North ER, Thomas S: An anatomic guide for arthroscopic visualization of the wrist capsular ligaments. J Hand Surg 1988;13A:815-822.
Palmer AK: Fractures of the distal radius. In Green DP, ed. Operative Hand Surgery, 2nd Edition. New York, Churchill Livingstone, 1988, pp991-1026.
Palmer AK, Werner FW: The triangular fibrocartilage complex of the wrist - Anatomy and function. J Hand Surg 1981;6:153-162.
Roger B, Chaise F, Laval-Jeantet M: Analyse tomodensitomé triques des modifications structurales du carpe, après section du ligament annulaire anterieur. J Radiol 1985;66:693-697.
Skahen JR III, Palmer AK, Levinsohn EM, et al.: Magnetic resonance imaging of the triangular fibrocartilage complex. 1990;15A:552-557.
Taleisnik J, Gelberman RH, Miller BW, Szabo RM: The extensor retinaculum of the wrist. J Hand Surg 1984;9A:495-501.
Taleisnik J: Wrist: anatomy, function, and injury. In Instructional Course Lectures, American Academy Orthopedic Surgeons. St. Louis, CV Mosby Co, 1978, pp61-87.
Taleisnik J: The ligaments of the wrist. J Hand Surg 1976;1:110-118.
Testut L: Traite d'anatomie humaine. Paris, Gaston Doin and Company, 1928, pp628-630.
Tinkelenberg J: Graphic reconstruction: microanatomy with a pencil. J Audiov Media Med 1979;2:102-106.
Weber ER: Concepts governing the rotational shift of the intercalated segment of the carpus. Orthop Clin North Am 1984;15:193-207.

Chapter 2

Kinematic Analysis

K-N. An and E.Y-S. Chao

Introduction

The main functions of the musculoskeletal system are to provide mobility and sustain load. It is reasonable to assume that mechanical factors play an important role in the function of this system. To determine these mechanical factors, motions and loads must be quantitated in precise mechanical terms. Kinematics is the study of body motion without reference to the forces causing this motion. This fundamental branch of dynamics finds a challenging application in the study of human movement.

Understanding the kinematics of human movement has both basic and clinically applicable applied value in medicine and biology. Motion measurement can be used to evaluate functional performance of limbs under normal and abnormal conditions. Kinematic knowledge is also essential for proper diagnosis and surgical treatment of joint disease and the design of prosthetic devices to restore function.

In kinematic analysis, a complete and accurate quantitative description of even the simplest movement requires large volumes of data and variables. These include position vectors, linear velocity and acceleration of the segment's center of mass, angular orientation, angular velocity, and angular acceleration of the segment in two planes. In order to describe these kinematic variables, a convention or coordinate system is required. Using a non-moving inertial system, the absolute motion of the segment can be described. On the other hand, with the local coordinates attached to each segment, relative motion between segments can be calculated. Either an absolute or relative description of body movement has an important role in the kinematic analysis of human movement. In this chapter, the basic analytic description and experimental methods available for kinematic analyses are presented.

Kinematic Models

When motion of an anatomic joint is to be measured a kinematic model of the joint should be established. Joint function is determined primarily by the shape and contour of the contact surfaces and constraints of the surrounding soft tissue. Global kinematics of the wrist is an interaction and accumulation of partial motions occurring at different levels within the joint.

In reality, all anatomic joints have six degrees of freedom (DOF) in which six independent parameters must be measured and described if the relative positions of the attached body segments are to be defined. From the engineering kinematic point of view, the classification of models depends on the degrees of freedom of motion. From the clinical perspective, medical classifications are usually based on the shape of two or more interactive joint surfaces.

Hinge Joint

A hinge joint, or revolute joint, is the simplest but most common model used to simulate an anatomical joint in planar motion. Movement of the moving segment is confined to one plane about a single axis embedded in the fixed segment. In general, hinge motion include both ginglymoid movement, such as in the elbow joint and the interphalangeal joints of the finger, as well as pivotal, or trochoid movement, such as the articulation between the radius and ulna where an arch-shaped surface rotates about a rounded, or peg-like pivot.

General Planar Joint

This model is usually used to simulate more general planar joint movement in which relative motion between all points takes place in parallel planes without a single fixed axis or center of rotation. This joint has three DOF, namely, two translations and one rotation. Usually, this motion consists of gliding movement such as that between the carpal bones of the wrist. In addition, this general three degrees of freedom planar-joint model has also been used for the analysis of knee-joint motion.

Universal Joint

The universal joint allows rotation about two axes through the joint and has two degrees of freedom. Anatomically, the universal joint is associated with two types of articulating joints. In one type, a saddle-shaped bone fits onto a socket that is concave-convex in opposite directions such as the articulation of the first metacarpal and trapezium. The other type of articulation is the ellipsoidal or condyloid joint such as the wrist joint where movement takes place in two perpendicular planes.

Ball and Socket Joint

This joint consists of a ball-shaped head that fits into a concave socket where movement of the moving segment takes place through rotation about three axes that intersect at the joint center. The relative motion is characterized by all points of the moving segment traveling on concentric spheres about the single joint center on a fixed body. A ball and socket joint, or enarthrosis, has three DOF and is the joint most commonly used to model three-dimensional joint movement such as that of the shoulder and hip joints.

General Spatial Joint

For completeness, a general spatial joint is included which does not assume any limitation on the number of degrees of freedom between the moving and fixed segments. The moving body is allowed six DOF, namely, three translations and three rotations. This model is commonly used when only detailed relative movement of the articulating surfaces is examined. The measurement and description of the relative motion associated with the general spatial joint are usually more complicated than are those of the simplified models. Nevertheless, it has been widely used for study of the human knee and wrist joints.

Analytic Description

Planar Motion

In general planar motion, the moving segments can have both translation and rotation about the fixed segment. Because of the translational component of motion, the center of rotation or axis of rotation for the moving segment will change throughout the course of motion. At any point in time, an approximate center of rotation can be determined, which is defined as the instantaneous center of rotation (ICR). Since the velocity of a point on a rigid body experiencing rotation during a small period of time must be perpendicular to a line joining the point and the center of rotation, this specific property can thus be used to determine the ICR graphically. In experimental measurements, however, it is nearly impossible to determine the velocity of different points on a body in motion. An alternate method for approximating the ICR was described by Franz Reuleaux in 1876 (Reuleaux 1963). In this method, the instantaneous locations of two points on the moving segment are identified from two consecutive positions within a short period of time, and the intersection of the bisectors of the lines joining the same points at the two positions defines the ICR (Figure 1). For a true hinged motion, the ICR will be a fixed point throughout the movement. Otherwise, loci of the ICR or centrodes will result. If the fixed segment is taken as the reference member and the motion of the moving segment is used to define the ICR, the resulting ICR will form a curve on the fixed segment called the fixed centrode. On the other hand, if the moving segment is used as the reference, the curve of the ICR determined by the relative motion of the fixed segment will form the moving centrode on the moving segment.

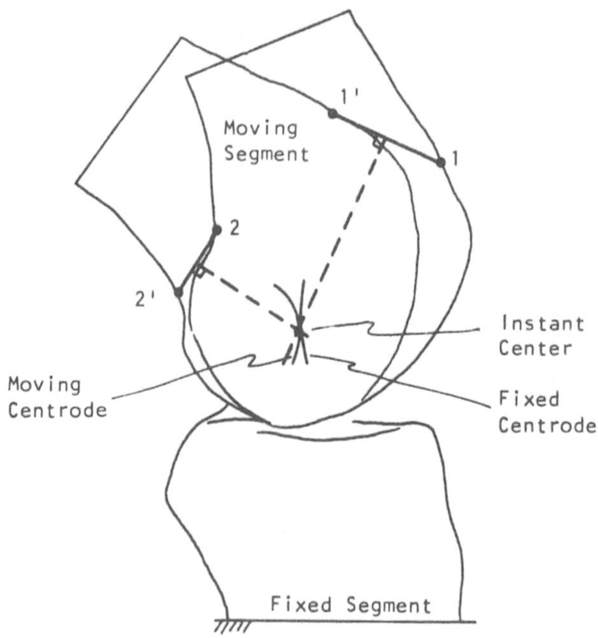

FIGURE 1. Determination of the instant center of rotation by Reuleaux's method. The points 1 and 2 displaced to 1' and 2', respectively, during the rotation of the moving segment. Perpendicular bisectors of these displacement lines intersect at the instant center for the displacement.

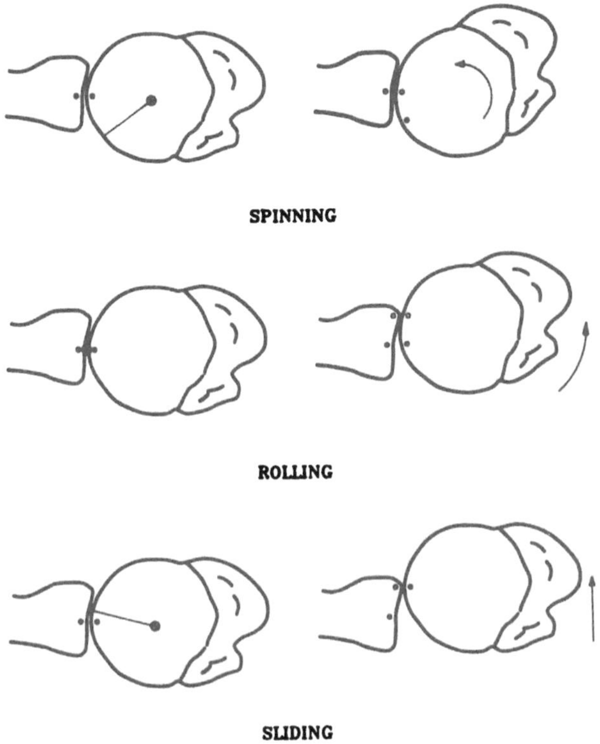

FIGURE 2. Three types of motion: spinning, sliding and rolling.

In practice, determination of the ICR by the above-described method is highly sensitive to error in the location of the points used to define the individual ICRs. The ratio of these errors increases exponentially as the individual displacements or time increments are made smaller (Panjabi 1979). Increasing the size of these incremented intervals will definitely decrease the error in determining the ICR. However, with larger intervals, the true kinematics will not be faithfully imitated.

For descriptions of general planar or gliding motion of the articular surfaces, the terms sliding, spinning and rolling are commonly used (Figure 2). Sliding motion is defined as the pure translation of a moving segment against the surface of a fixed segment. The contact point of the moving segment does not change, while its mating surface has a constantly changing contact point. If the surface of the fixed segment is flat, the ICR is located at infinity; otherwise it will be located at the center of the curvature of the fixed surface. Spinning motion is the exact opposite of sliding motion, where the moving segment is, in this case, located at the center of the spinning body that is undergoing pure rotation. Rolling motion is motion between moving and fixing surface where the contact points on each surface are constantly changing. However, the arc length of the moving surface matches the path on the fixed surface so that the two surfaces have point-to-point contact without slippage. The relative motion of rolling is a combination of translation and rotation. The ICR is located at the contact point. Most of the planar motion of anatomical joints can be described using a combination of any two of the above three basic types of motion.

Eulerian Angle System

When describing motion of an anatomic joint, commonly only the rotations are considered. For general joint rotation in space, three angles are required. However, for finite spatial rotation, the sequence of rotation is extremely important and must be specified for a unique description of joint motion. However, with proper selection and definition of the axes of rotation between two bony segments, it is possible to make the finite rotation sequence independent or commutative (Chao 1980; Suntay, et al., 1978). In this selection of axes, one is fixed to the fixed segment and another is fixed to the moving segment. In the knee joint, for example, the flexion-extension angle (ϕ) occurs about a mediolaterally directed axis fixed to the femoral condyle, and axial rotation (ψ) is measured about an axis along the shaft of the tibia (Figure 3). The third axis (also defined as the floating axis) is orthogonal to them and defines abduction-adduction (θ). These rotations match Eulerian angle description.

If a unit vector triad (I, J, K) is fixed to the fixed segment along X, Y, Z axes and another triad (i, j, k) is fixed to the moving segment, along x, y, z axes (Figure 3), the relationship between them after any arbitrary finite rotation can be expressed by a rotational matrix in terms of the Eulerian angles (ϕ, θ, ψ).

$$\begin{vmatrix} \underline{i} \\ \underline{j} \\ \underline{k} \end{vmatrix} = \begin{vmatrix} c\phi c\theta & s\phi c\theta & -s\theta \\ -s\phi c\psi + c\phi s\theta s\psi & c\phi c\psi + s\phi s\theta s\psi & c\theta s\psi \\ s\phi s\psi + c\psi s\theta c\psi & -c\phi s\psi + s\psi s\theta c\psi & c\theta c\psi \end{vmatrix} \begin{vmatrix} \underline{I} \\ \underline{J} \\ \underline{K} \end{vmatrix} \qquad (1)$$

where s and c stand for sine and cosine, respectively. The Eulerian angles can be calculated based on the known orientation of these unit vector triads attached to the segments.

The advantage of this system for description of spatial rotation of anatomic joints is that the angular rotations do not have to refer back to the neutral position of the joint because the rotational sequence can be totally independent and the measurement can be easily related to anatomic structures. However, it is important to recognize that two of the rotational axes in this system are non-orthogonal when the joint departs from its neutral position. Consequently, the system is difficult to use in kinetic analysis. Angular velocity and acceleration have to be transformed into a set of inertial axes in terms of the Eulerian angles defined.

Generalized Six-DOF Joint Motion

The most commonly used analytic method for the description of six DOF spatial motion is a screw or a helic displacement axis (SDA) (Kinzel, et al., 1972; Spoor and Veldpaus, 1980; Youm and Yoon, 1979).

This concept states that the relative displacement of a moving segment from one position to another can be defined in terms of a rotation (ϕ) about and a translation (t) along a unique axis called the screw displacement axis which is located within the fixed segment (Figure 4). The advantage of using a screw axis is that the orientation of the SDA remains invariant, regardless of the reference coordinate axes used. The screw displacement axis is a true vector quantity; its magnitude can be decomposed along any coordinate axes used for analysis. However, the amount of the finite screw rotation that is not a vector quantity and the decomposition of it must be carefully interpreted because of the noncommutative nature of finite rotation.

The concept of SDA has been applied to the study of the kinematics of the elbow joint to demonstrate that this joint follows a tight hinge axis motion pattern. Similarly, the system has been used to show that the shoulder joint follows the ball-and -socket joint assumption. Documentation of carpal bone motion in the wrist has also been described by utilizing this SDA concept.

Numerous methods can be used to determine the orientation and location of the SDA (Kinzel, et al, 1972; Spoor and Veldpaus, 1980; Youm and Yoon, 1979). The SDA is parallel to the angular velocity vector; therefore, the orientation of the SDA can easily be defined if the angular velocity of the moving segment is available. However, since the velocity is expressed instantaneously, the precise location of the SDA can be difficult to locate within the displacement of a finite time period. In addition, determination of linear and angular velocities represents a significant experimental problem.

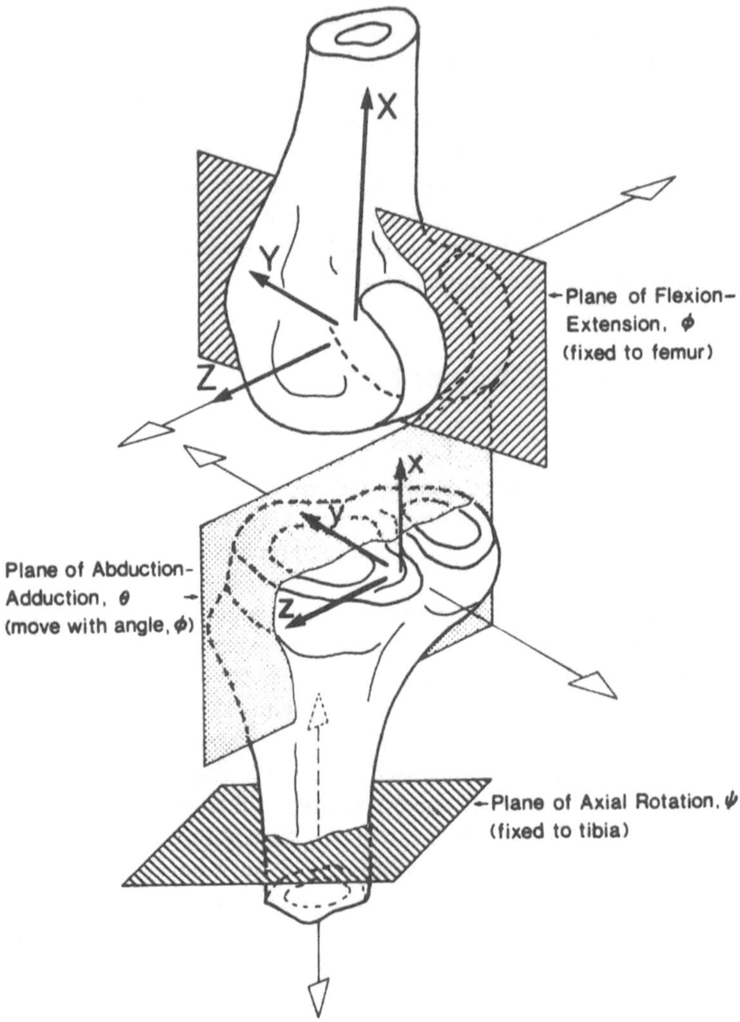

FIGURE 3. Description of the knee joint motion by Eulerian angle system.

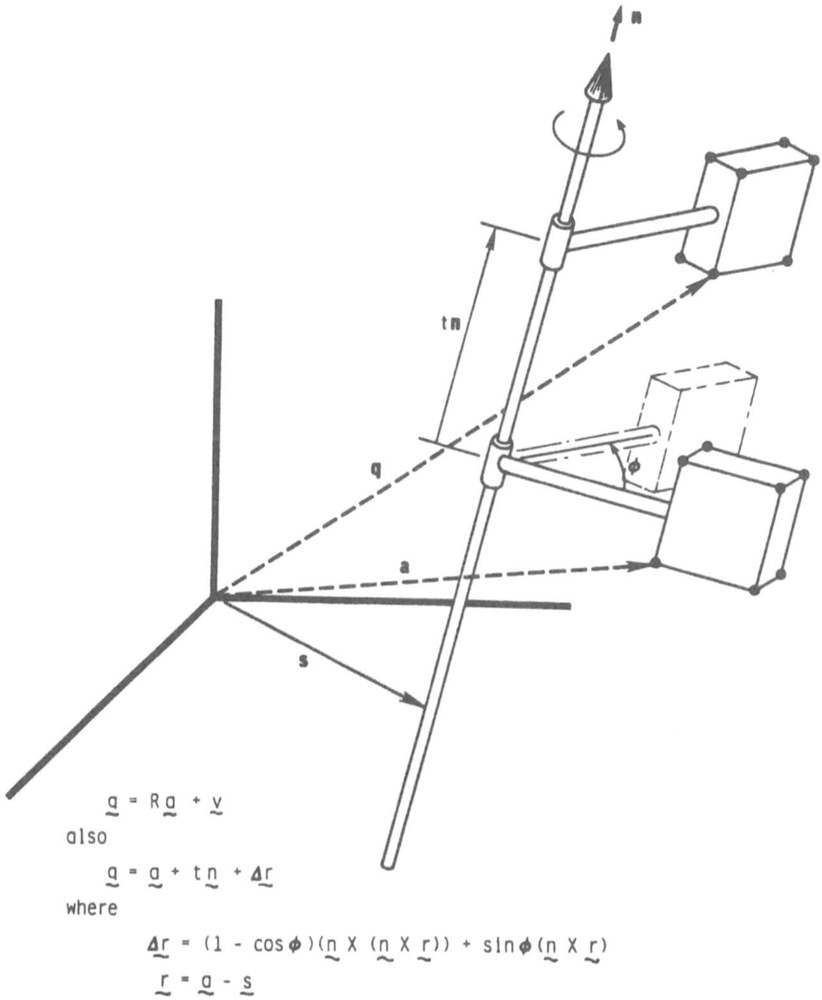

FIGURE 4. Description of the generalized 6-DOF joint motion by using screw displacement axis. The relative displacement of a moving segment from one position to another can be defined in terms of a rotation (ϕ) about and a translation (t) along a unique screw displacement axis.

An alternate method to determine the orientation of the SDA and screw rotation is based on a rotational matrix used to define the location of reference points with respect to a coordinate system after finite rotation. This rotational matrix can be expressed in terms of directional cosines between the coordinate axes before and after rotation (Goldstein 1959). The screw rotation ϕ may be found from the trace of the rotational matrix:

$$\phi = \cos^{-1} \{ (tr[R] - 1)/2 \} .$$

With ϕ known, the components for the orientation of the screw axis can be calculated. From the translation vector, which can be obtained by comparing the position vectors of the reference points on the moving body, the amount of translation along the SDA and the position vector of a point on the SDA can be calculated.

Measurement System

Numerous experimental methods to measure the kinematics of human movement have been developed in the past years. In nearly all methods, basic assumptions have been made in order to facilitate analytic description because of the complexity of the system involved. Although the precise instruments and techniques involved are different, they generally follow similar principles, and each has its unique advantages and disadvantages.

Electromechanical Linkage Methods

These methods use exoskeletal linkage systems containing rotatory potentiometers, called electrogoniometers. The linkages are fastened to both the proximal and distal limb segments which constitute the moving and fixed bodies between which the relative motion is measured (Figure 5). Depending upon the sophistication of the mechanism design, the electromechanical linkage system can be used to measure simple hinge joint motion, three-dimensional angular rotation, or generalized 6-DOF rigid body relative motion (Chao 1980; Chao, et al., 1980).

There are several advantages to using the linkage system. It is easy to use and can instantaneously provide direct measurement of joint relative motion, without a tedious data reduction process. It is reliable and reproducible, and the accuracy is generally acceptable, particularly for clinical application. However, there are several inherent limitations. First of all, the linkage system usually provides relative motion which cannot be used directly for kinetic analysis. Instantaneous monitoring of the motion of one segment with respect to the inertial reference frame is necessary in order to achieve this purpose. Secondly, application of the linkage system is usually not located within the joint, so any misalignment will cause significant cross-talk among the potentiometer readings and will make the instantaneous reading associated with the anatomical description difficult. However, such cross-talk error can be theoretically corrected (Chao 1980).

Stereometric Methods

When three noncolinear points fixed to a rigid body are defined within an inertial reference frame, the position and orientation of that rigid body can be specified, and the relative rotation and translation occurring at a joint can be determined. The general principle of the stereometric method used for kinematic measurement is, basically, to monitor the locations of these reference points.

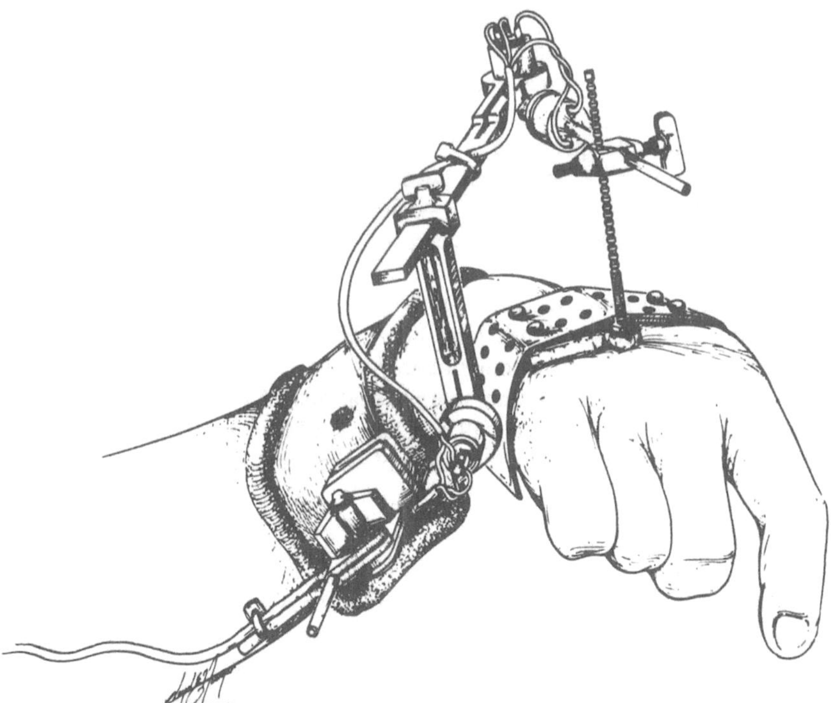

FIGURE 5. Triaxial electrogoniometer for wrist joint motion measurement. Flexion-extension, abduction-adduction motion can be monitored simultaneously.

The reference points on the body can be conveniently represented by markers in the form of miniature light bulbs, reflecting dots, light emitting diodes (LED), and ultrasonic transducers. There are several systems used to monitor and define these reference points. The stereophotographic system is the simplest and uses two or three still cameras or high-speed movie cameras located at various orientations. The spatial locations of these reference markers can then be constructed based on the film data viewed from oblique angles. Based on the same principle, TV cameras have also been used. The data on these reference locations can be used for not only the calculation of relative motion of the limbs at the joint but also the absolute motion with the inertial reference frame. The major disadvantage, of course, is the tedious procedures involved in data reduction and analysis. Parallax and magnification errors should be noted and compensated for analytically based on the optical principle (Shapiro 1978).

More recently, with the development of analog photodetectors or image detectors in sensing units of standard TV camera lenses, location of the reference points by using either light reflectors or an LED can be automatically monitored and scanned by computer. These devices make the data reduction process much easier.

Based on a similar principle, a technique utilizing hypersonic impulses to represent the reference points was developed. Instead of using a camera, the spatial positions of the sonic transducers are directly monitored by microphone sensors and a microprocessor device (Youm and Yoon, 1979).

Finally, in the same category as the stereometric method, the biplanar roentgenographic method has long been used to identify bony reference points for more accurate movement analysis of the skeletal system. Knowing the X-ray tube focal point, the spatial locations of the marker points are uniquely defined (Brown, et al., 1976; Chao and Morrey, 1978). This method has been used extensively for the study of wrist kinematics.

Magnetic Tracking

A magnetic position and orientation tracking system has been developed which can determine the three dimensional position and orientation of a sensor relative to a source in space (Raab, et al., 1979). The six-degrees of freedom measurements are accomplished by using low-frequency magnetic field technology to interpret the interaction of magnetic field between three sets of orthogal coils contained in both the source and sensor. Various commands are available to set parameters values of the system and to obtain information and data from the system. With appropriate alignment and attachment of the source and sensors to anatomical structures, the relative location and orientation of anatomical elements can be monitored (An, et al., 1988).

Imaging Techniques

Imaging technologies, such as computerized tomography (CT) and magnetic regruance imaging (MRI) have been rapidly developed in the past decade, facilitating the development of in vivo investigation of human skeleton kinematics. However, due to difficulties in consistent identification of markers of bony segments, most studies using these technologies have not been quantitative in nature. Applications of feature extraction techniques to digital models provided a novel approach to the needed quantative definition of bony reference system (Belsole, et al., 1988). The application of algebraic analysis of the CT digital data provided mathematical markers to establish the principal axes of individual carpal bones. Once this was done the position and orientation of that bone could be calculated at various wrist positions.

Summary

Kinematic analysis of human movement has both basic and applied value in medicine and biology. Gross measurement of motion can be used as a tool for evaluation of functional performance of limbs under various surgical or therapeutic treatments. Kinematic knowledge of precise articular surface movement is essential in the design or prosthetic devices for the restoration of joint function. Various levels of sophistication in terms of theoretical modeling and experimental measurement techniques are available. The selection of each of these techniques depends on the objective of the study. Above all, the advantages, disadvantages and limitations associated with each of these techniques should be understood.

References

An KN, Jacobsen MC, Berglund LJ, Chao EYS: Application of a Magnetic Tracking Device to Kinesiology Studies. J. Biomechanics 1988;21:613-620.

Belsole RJ, Hilbelink DR, Llewellyn JA, Stenzler S, Green TL, Dale, M: Mathematical Analysis of Computerized Carpal Models. J. Orthop Res. 1988;6:116-122.

Brown RH, Burstein AH, Nash CL, Schock CC: Spinal analysis using a three dimensional radiographic technique. J. Biomech. 1976;9:355-365.

Chao EYS: Justification of triaxial goniometer for the measurement of joint rotation. J. Biomech. 1980;13:989-1006.

Chao EYS: Experimental methods for biomechanical measurements of joint kinematics. In: CRC Handbook of Engineering in Medicine and Biology, Section B: Instruments and Measurements, edited by BN Feinberg and DG Fleming. West Palm Beach, Florida: CRC Press 1978;pp385-411.

Chao EYS, Morrey BF: Three-dimensional rotation of the elbow. J. Biomech. 1978;11:57-73.

Goldstein, H: Classical Mechanics. Reading, Massachusetts: Addison Wesley Publishing Co., Inc. 1959;pp107-124.

Kinzel GL, Hall AS, Hillberry BM: Measurement of the total motion between two body segments, Part I-Analytic development. J. Biomech. 1972;5:93-105.

Panjabi MM: Center and angles of rotation of body joints: A study of errors and optimization. J. Biomech. 1979;12:911-920.

Raab FH, Blood FB, Steiner TD, Dones HR: Magnetic position and orientation tracking system. IEEE Trans. Aero Space Elector System. AES 15;709-718.

Reuleaux F: The Kinematics of Machinery: Outline of a Theory of Machines. London: Macmillan, 1876. Translated by ABW Kennedy, New York, New York: Dover Publications, Inc. 1963;p61.

Shapiro R: Direct linear transformation method for three-dimensional cinematography. Res. Q. 1978;49:197-205.

Spoor CW, Veldpaus FE: Rigid body motion calculated from spacial coordinates of markers. J. Biomech. 1980;13:391-393.

Suntay WJ, et al.: A coordinate system for describing joint position. Advances in Bioengineering. ASME 1978;pp59-62.

Trocme MC, Sather AH, An KN: A biplanar cephalometric stereo radiography technique. Am. J. Orthod. Denfofac. Orthop. 1990;98:168-175.

Youm Y, Yoon YS: Analytical development in investigation of wrist kinematics. J Biomech 1979;12:613-621.

Chapter 3

Wrist Joint Motion

J. Ryu, A.K. Palmer, and W.P. Cooney, III

Introduction

Kinematics, that branch of dynamics that deals with motion apart from considerations of mass and force, has been grafted by surgeons into that branch of treatment attempting to return injured joints to functional parameters in accord with range of motion required for performance of patient activities. An injured wrist presented to hand surgeons for repair demands a careful interweaving of strength, dexterity and stability to make that human hand useful. The mending of such a complex structure is, however, a formidable challenge. When presented with maladies of the wrist, many well intended remedies have been designed and promoted, but few have been tested to reveal true efficacy.

Part of this dilemma stems from the fact that little study has been devoted to defining the normal functional ranges of wrist joint motion. How can an injured wrist be returned to normal function, if its functional capacity has yet to be truly appreciated?

Although normal maximum motion of the wrist has been previously documented using standard hand goniometry, (American Academy of Orthopaedic Surgeons, 1965; Boone and Azen, 1979; Kraft and Detels, 1972; Pryce 1980; Sarrofian, et al., 1977; Volz, et al., 1980) only recently has instrumentation been developed to assess normal wrist motion during activity. Brumfield et al. (1966) obtained data on wrist flexion/extension arcs using a uniaxial goniometer, while Palmer et al. (1985) analyzed wrist motion in ten normal subjects with a polyaxial electrogoniometer.

Considerable interest in functional range of motion following partial wrist fusion (Kleinman, et al., 1982; Speltz, et al., 1983; Watson and Hempton, 1980; Watson, et al., 1985), total wrist arthroplasty (Cooney, et al., 1984) and wrist ligamentous reconstruction (Palmer, et al., 1978) has been shown, but these are of limited use until the issue of range of motion requirements for daily living within "normal" populations is adequately addressed.

The aim of this chapter is to describe the range of wrist motion involved in the performance of certain daily living activities. If this information is used as a data base for comparison with subjects suffering wrist abnormalities, and for the assessment of wrist reconstructive procedure functional outcome, it will have served its purpose.

Methodology

Several techniques of studying overall wrist motion, including stereometric, accelerometric and electromechanical linkage have been described. Stereometry using skin markers (placed to locate the small, underlying wrist bones) has been felt to lack sensitivity when detecting biplanar changes in wrist motion (Ayoub, et al., 1970; Engen and Spencer, 1968; DeRoos, et al., 1977; Eberhart and Inman, 1967; Erdman, et al., 1976; Yoon 1979). The accelerometric technique requires complex data reduction, which may be excessive when only joint kinematic data is desired (Hayes, et al., 1978; Morris 1973).

The multiaxial electrogoniometer, which provides an of electromechanical linkage avoids a number of these pitfalls (Palmer, et al., 1985; An and Chao, 1984; Chao 1980; Chao 1978; Chao and Morrey, 1978). It offers ease of application, real-time analog data, describing multiplanar joint motion, and experimental reliability and reproducibility with acceptable degrees of accuracy (An and Chao, 1984; Chao 1980; Chao 1978; Karpovich and Karpovich, 1959; Morrey, et al., 1981). The axes of electrogoniometer rotation can be made co-linear with the estimated axes of wrist flexion/extension and radial/ulnar deviation landmarks on the hand and wrist. Alignment can be confirmed via biplanar x-rays. Although the goniometer linkage system carries a certain amount of inherent inertia, standard daily living activities seem to involve slow hand and forearm movement. Thus, motion artifact should be minimal.

An example of a wrist electrogoniometer used in one recent study is shown in Figure 1. This particular design allows application to either upper extremity. It is attached to each subject with two universal mounting components, one attached to the dorsal part of the hand (metacarpal component) and the other to the distal forearm (radial component) (Figures 2a and 2b). Its lightweight, low-profile design reduced the effects of inertia while avoiding interfering with normal wrist motion during performance. The wrist joint was assumed to have two degrees of freedom, namely flexion/extension and radioulnar deviation. This biaxial electrogoniometer can thus be designed to record motion within such planes. To gather data from both planes simultaneously, two electropotentiometers were attached to the electrogoniometer such that they were mutually perpendicular to each other, remaining so throughout wrist motion. The linkage between potentiometers and the mounting apparatus employed geometric parallelogram principles. Motion observed was recorded directly on a model 5314 strip chart (Soltec Corp., Sun Valley, CA), allowing exact degrees of wrist motion to be hand reduced and calculated.

Alignment of the goniometer to the wrist is essential for accurate recording of the range of wrist motion required for the performance of daily living activities. For purposes of this investigation, the head of the capitate nearest the lunate contact area was assumed to represent the center of wrist rotation in both flexion/extension motion and radioulnar deviation.

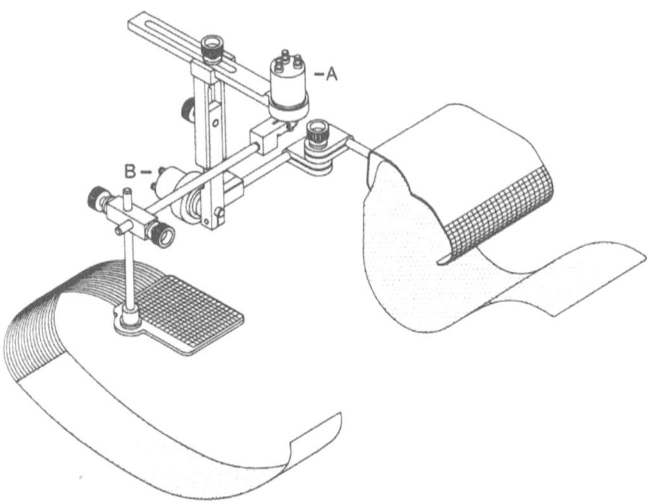

FIGURE 1. Schematic diagram of the biaxial electrogoniometer with 2° of freedom for wrist motion measurement during activities of daily living. A) Dorsal potentiometer; B) Ulnar potentiometer.

To assure reliable and reproducible placement of the goniometer on test subjects, cadaver studies were carried out in which a T-marker was placed on the dorsal intercarpal area of the wrist, complemented by indelible marks on the skin (Figures 3a, 3b and 3c). Tests in five cadaveric hands pointed to the medial cortex of the third metacarpal, the volar aspect of the pisiform, the base of the first metacarpal and the ulnar head as accurate, accessible anatomical landmarks.

The axis of radioulnar deviation was identified by drawing a line along the ulnar border of the third metacarpal upon which the mid point between the distal radius and the proximal end of the third metacarpal was marked. To identify location of the flexion/extension axis, this same point was extended ulnarly, perpendicular to the forearm bones until it reached the midpoint between the distal radius and the pisiform's volar edge.

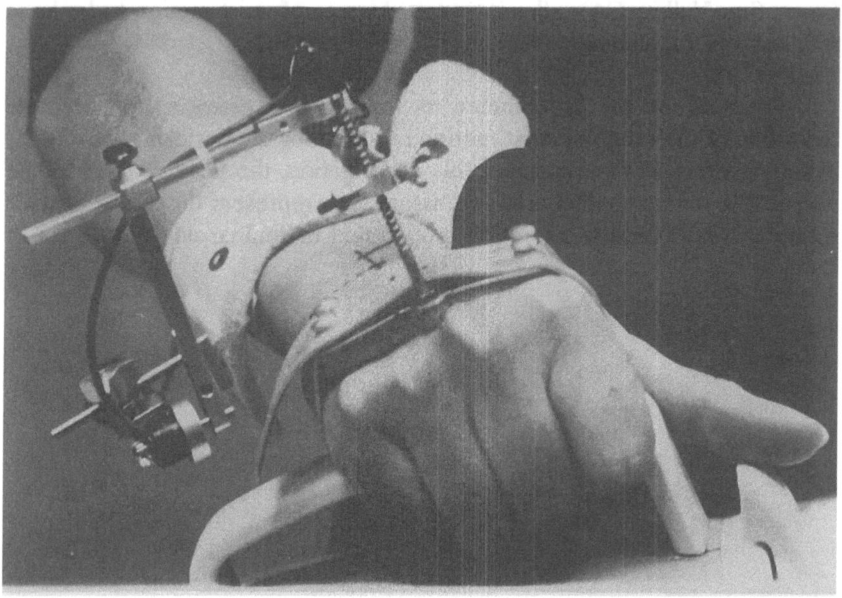

FIGURE 2(a). From a distal to proximal view of the hand and wrist, alignment of the dorsal potentiometer (radioulnar deviation axis) is made with respect to the third metacarpal, and alignment of the ulnar potentiometer (flexion-extension axis) is made with the midposition of the forearm from previously determined skin markers.

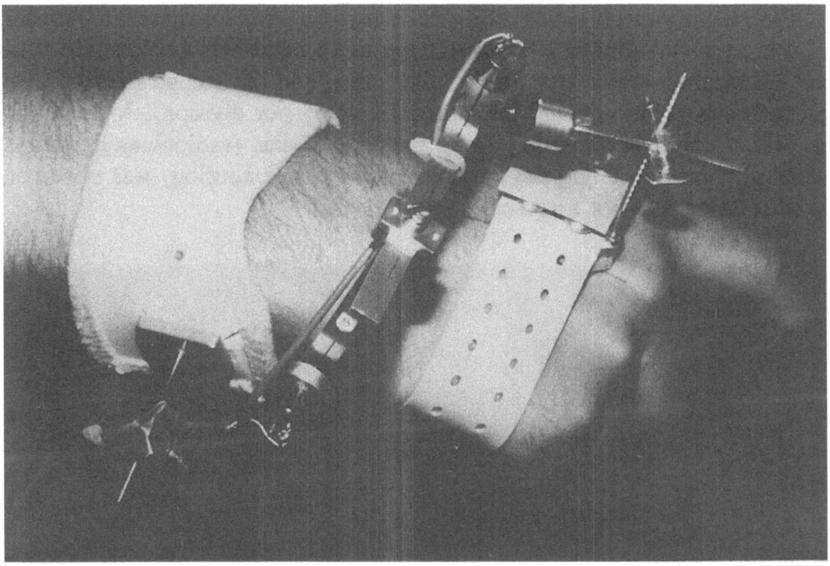

FIGURE 2(b). Dorsal view of the wrist, goniometer placement. Support straps attach the distal goniometer pad to the hand with alignment along the medial border of the third metacarpal. The dorsal potentiometer is centered over the head of the capitate. The position of the potentiometer can be adjusted through the set screw attached to the cross-bars. An orthoplast cuff and metal support secure the goniometer to the distal forearm. The ulnar potentiometer is positioned on the ulnar side of the wrist and is also centered in line with the head of the capitate. The placement of the ulnar potentiometer can be adjusted with the set screw attached to the wrist cuff.

Open dissection of the wrist accompanied by radiographic studies were performed in the cadavers to validate T-marker placement. To further ensure alignment, T-markers were also placed on the wrist of three live subjects, and biplanar (AP-lateral) x-rays were taken to determine whether or not markers were in line with the ulnar border of the third metacarpal and that the cross-bar approximated the head of the capitate. All these tests indicated that goniometer placement could be accurately performed such that alignment coincided with wrist flexion/extension and radioulnar deviation using easily palpable anatomic landmarks.

Electrogoniometer calibration was performed using a strip chart recorder, a constant power source, a fixation frame and a hand-held goniometer. With one electrogoniometer arm fixed to the frame, and the second free to move, the zero axis position could be chosen (scale 5° per division). A total 110° of motion could be recorded along the sagittal axis (radioulnar deviation) while 180° motion along the frontal axis (flexion/extension) was observed. Axial rotation was not assessed.

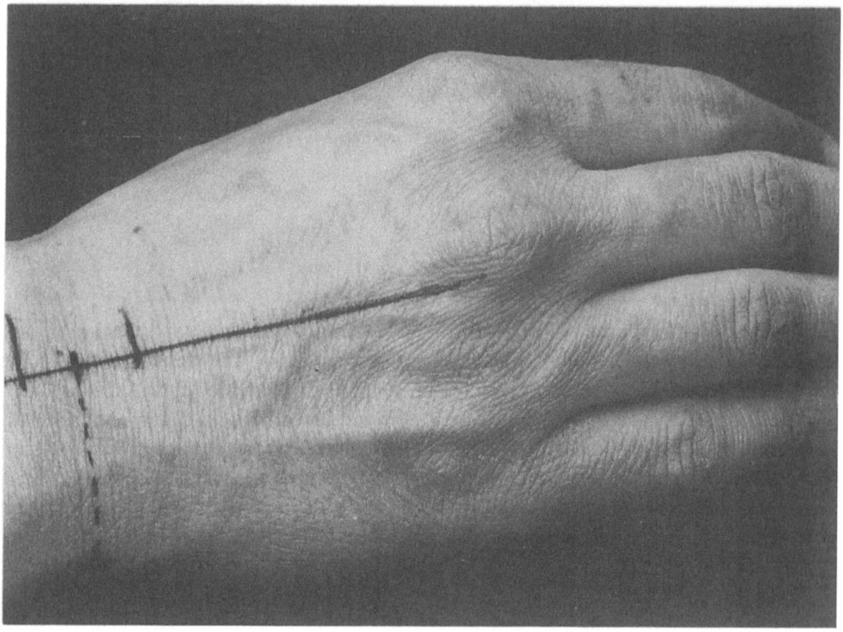

FIGURE 3(a). The dorsal potentiometer is aligned over the center of radioulnar rotation which is determined by extending a straight line along the ulnar border of the third metacarpal to the distal radius. This line is marked at the distal radius and the proximal end of the metacarpal. The axis of motion is taken as the midpoint between the two marks on the line.

The ulnar potentiometer is aligned to on the ulnar side of the wrist at the midpoint between the volar aspect of the pisiform and dorsal aspect of the ulnar head. The goniometer is then centered on the dotted line (above) which extends perpendicular from the third metacarpal line. the wrist is maintained in a neutral position of flexion/extension and radioulnar deviation throughout the alignment procedure.

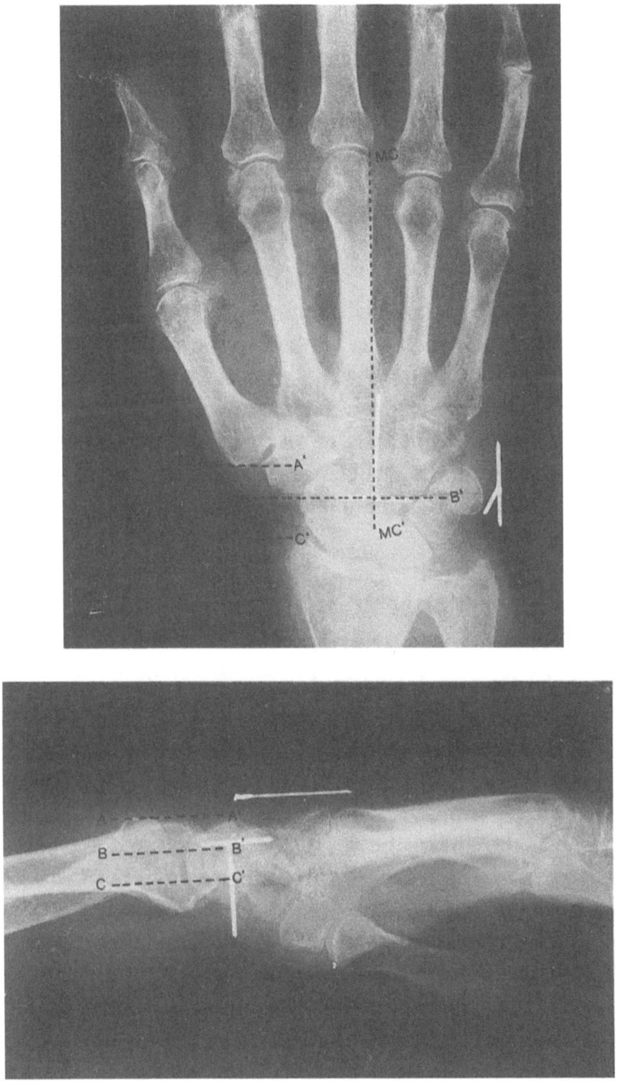

FIGURE 3(b) and 3(c). 3(b) AP view of T-markers; 3(c) lateral view of T-markers. Based on anatomic landmarks (see test), the dorsal and ulnar potentiometer placement is confirmed in cadavers by radiographs of the wrist with T-markers (entered) on the radioulnar deviation axis (AP radiograph) and flexion-extension axis (lateral radiograph). In Figure 3(b), line A-A' is base of the first metacarpal; line B-B' is midpoint of the pisiform; line C-C' is the tip of the radial styloid. MC-MC' is medial border of the third metacarpal. In Figure 3(c), line D-D' is dorsal aspect of the ulnar head. E-E' is volar surface of the pisiform and line F-F' is the midpoint between the ulnar head and pisiform.

Forty normal subjects with no history of upper extremity injury or pathology and in good health were tested. The subjects expressed a wide range of heights (5' to 6'4", average 5'4") and weights (115 to 240 lbs., average 160 lbs.) They were divided into four groups of ten: ten males and ten females between 20 and 39 years of age, and ten males and ten females between 40 and 60 years of age. Each underwent thorough examination to rule out wrist-joint pathology before goniometry testing. The electrogoniometer was placed on the dominant right hand in all forty subjects.

In a second study, a triaxial goniometer was developed to analyze functional wrist motion which was applied to the upper extremity by the use of orthoplast brackets and tape. The results from this study report data derived from the flexion/extension and radial/ulnar deviation planes of motion only. Technical difficulties in securing the devise adequately to the subjects skin prohibited the collection of accurate and reproducible data regarding carpal pronation and supination (Palmer, et al., 1985).

The Analytical Approach

In the study by Ryu et al. (1991) thirty-one activities reflecting the major functional requirements of the wrist were analyzed. They were intended to represent standard activities of daily living, allowing for assessment of distal upper extremity function. The first set of activities was comprised of seven "palm placement" positions: on top of the head, the back of the head, the front of the neck, the chest, the waist, the sacrum, and the right foot. The second category involved personal care and hygiene, the third consisted of food preparation motions, while the fourth consisted of typical work functions, (hammering, turning a key, grasping a door knob, etc...) No restriction of wrist motion could be attributed to the electrogoniometer throughout these activities except for that of lifting oneself up from a chair, during which extreme extension seemed to be compromised. Even in this situation, consistent results were obtained.

The study was performed with the subjects either sitting upright in an adjustable chair, or standing upright. Subjects were allowed to adjust their positions to those of maximum comfort. After a short description of each activity was given verbally, each subject was asked to perform the activity in a manner to which they were accustomed (activities were not demonstrated for them.) Each activity was repeated three times. Flexion and extension were measured with the forearm in neutral rotation, while radial and ulnar deviation were measured with the forearm pronated and the palm placed on a flat surface to limit flexion and extension. Mean values were values were used as representative data.

Motion patterns for each activity's repetitive trials were recorded. This procedure tests the ability of the individual to reproduce motion patterns for each specific activity and demonstrated system reliability (Figure 4). Extreme motions in all four directions (flexion, extension, radial and ulnar deviation) were recorded at the beginning, between each of the four activity categories, and at the end of each test battery to detect any goniometer movement in relation to the forearm and hand. These measurements were compared with hand-held goniometer measurements in a test of instrument reliability (Figure 4). Total motion arc in both flexion/extension and radioulnar deviation coincided extremely well between the hand-held goniometer and electrogoniometer readings (mean difference >2°) (Table 2).

The relative amount of motion registered by the electrogoniometer depended on the definition of neutral (zero) wrist positioning. For the purpose of this study, the zero position was established by placing the hand on a flat surface, leaving it in a slightly flexed, slightly radially deviated position.

Data reduction was initially performed for each group of ten. Ranges of activity motion for each of the four groups were similar. Non-parametric analysis (Kruskal Wallis H Test) was use to determine this as it was unlikely the data would follow patterns of normal distribution. On the twenty-four motion activities and seven placement functions, results of tests on all four data sets (flexion, extension, radial and ulnar deviation) were compared. No significant differences were found between similar data sets among the four subject groups. Data degree fractions were rounded off to simplify result presentation.

In the study by Palmer et al. (1985), the triaxial electrogoniometer was used to study the functional wrist motion of ten normal human subjects during activities of daily living. Twenty-four standardized tasks that simulated personal hygiene, culinary skills, and miscellaneous activities of daily living were performed as wrist motion was simultaneously analyzed in three axes. Data was stored and reduced using a PDP 11/23 minicomputer. Additionally, specific tasks related to location were evaluated, including seven tasks related to carpentry, five tasks related to housekeeping, five tasks related to mechanic work, four tasks related to secretarial work, and five tasks specific to a surgeon were evaluated in a similar manner. Each task was performed in a standardized manner and repeated three times providing an average to be used in a final data presentation (Table 1).

TABLE 1. Fifty-two standardized tasks broken down by activity.

Standard tasks (1 to 26)	Carpenter (C_1 to C_7)
1. ROM--F/E	C_1, Handsaw
2. ROM--R/U	C_2, Power saw
Personal hygiene	C_3, Power drill
3. Comb hair	C_4, Screwdriver
4. Touch tailbone	C_5, Ruler (unfold)
5. Wring washcloth	C_6, Hammer nail
6. Button shirt	C_7, Pull out nail
7. Tie shoe	Housekeeper (H_1 to H_5)
Culinary	H_1, Broom
8. Cut with knife	H_2, Dustpan
9. Eat with fork	H_3, Sponge
10. Turn spatula	H_4, Wastebasket
11. Drink from cup	H_5, Vacuum
12. Pour from pitcher	Mechanic (M_1 to M_5
13. Turn can opener	M_1, Socket wrench (vertical)
14. Stir in bowl	M_2, Socket wrench (horizontal)
15. Open/close jar	M_3, Wing nut
Other ADLs	M_4, Cotter pin
16. Open/close faucet	M_5, Mallet
17. Page through book	Secretary (SE_1 to SE_4)
18. Print name	SE_1, Sharpen pencil
19. Put phone to ear	SE_2, Answer phone
20. Dial "0"	SE_3, Write
21. Load typewriter	SE_4, Fold letter
22. Turn steering wheel	Surgeon (SU_1 to SU_5)
23. Turn doorknob	SU_1, Cut with scalpel
24. Turn key	SU_2, Suture
25. Turn screwdriver	SU_3, Square knot (instrument)
26. Throw ball	SU_4, Square knot (free hand)
	SU_5, Scissors

F/E = flexion/extension
R/U = radial/ulnar

TABLE 2. (A-C) Mean extreme range of wrist motion, measured by hand goniometer and electrogoniometer before and after subject study. (D) The electrogoniometer (first reading) is compared to the hand goniometer reading. (E) The first and last electrogoniometer reading after completing the test sequence are compared for each subject. The accuracy of the electrogoniometer is evident by the small difference in he range of Extension/flexion and Ulnar/Radial noted (D and E).

	Ext.	*Flex.*	*Range (Ext./ Flex.)*	*U.D.*	*R.D.*	*Range (U.R.)*
A. Hand goniometer	59.3	79.1	138.4	37.7	21.1	58.8
B. Electrogoniometer (first)	68.9	68.4	137.3	42.7	17.8	60.57
C. Electrogoniometer (last)	69.3	68.1	137.4	42.9	16.6	59.5
D. Electro (first)--hand	9.6	-10.7	-1.1	5.0	3.3	1.7
E. First--last electro	-0.4	0.3	0.1	-0.2	1.2	1.0

Legend: TABLE 2.

Ext. = Extension
Flex. = Flexion
U.D. = Ulnar Deviation
R.D. = Radial Deviation
U.R. = Ulnar/Radial Deviation Total

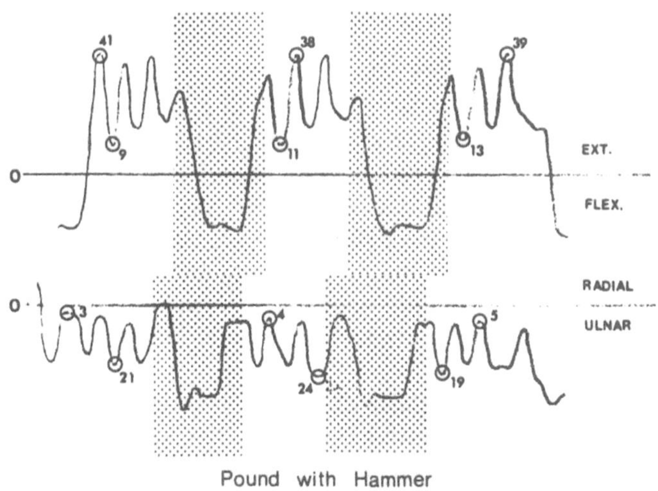

FIGURE 4(a). The range of motion which was required for each activity was assessed directly by strip chart recorder (Soltec Corp., model 3314). Motion from the rest position to the beginning of each activity (shaded areas) was discarded. In this example, three consecutive motions of the wrist involved in pounding with a hammer are recorded.

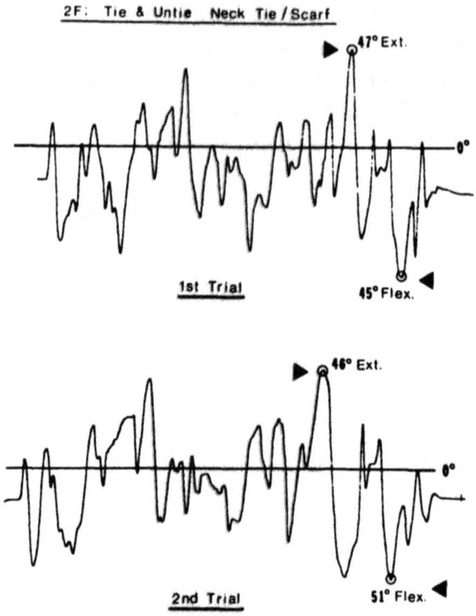

FIGURE 4(b). Reproducibility of the biaxial electrogoniometer is demonstrated in virtually all subjects and all activities, drawing very similar patters of motion (see arrows above) in multiple trials even in complex wrist activities such as tying and untying a neck tie.

Assessment of Wrist Motion

The results from the study by Ryu et al. (1991) revealed that the average maximum range of wrist motion was 59° extension to 79° flexion (138° arc of motion), and 21° radial deviation to 38° ulnar deviation (59° arc of motion). Comparison of these findings with those of previous studies (American Academy of Orthopaedic Surgeons, 1965; Boone and Azen, 1979; Brumfield, et al., 1966) showed general agreement, although full extension measurements were slightly smaller in the current study (Table 3).

In the studies of seven palm placement activities, mean and standard deviations of wrist placement were obtained (Table 4), (Figure 5). Placement of the palm on the head, neck, chest, waist, sacrum and foot, the wrist was in slight flexion (0° to 24°). A wide range of flexion and extension motions were required to perform these activities. In general, most were performed with the wrist somewhere between neutral and 20° flexion. More extension-oriented ranges were seen accompanying daily living activities requiring continuous motion. This agrees with the findings of Brumfield et al. (1966). Hand palm placement activities also required an arc of 5° radial to 15° ulnar deviation. Placing the hand on the sacrum involved the most extensive range of motion, 48° ulnar deviation.

TABLE 3. Comparison of extreme range of wrist motion in the study with previously reported data.

	Ext.	*Flex.*	*Range (Ext./ Flex.)*	*U.D.*	*R.D.*	*Range (U.R.)*
AAOS (1)	71	73	144	33	19	52
Boone et al. (5)	74.9	76.4	151.3	36.0	21.5	57.5
Brumfield et al. (7) Male	64	73	137			
Brumfield et al. (7) Female	65	82	147			
Palmer et al. (31)			133.3			40.5
This study	59.3	79.1	138.4	37.7	21.1	58.7

Legend: TABLE 3.

Ext. = Extension
Flex. = Flexion
U.D. = Ulnar Deviation
R.D. = Radial Deviation
U.R. = Ulnar/Radial Deviation Total

TABLE 4. Positions of the wrist during palm placement activities (mean and standard deviation).

	Extension and flexion (degrees)*	*Ulnar and Radial** Deviation (degrees)*
Top of the head	- 20.9 ± 13.9	16.1 ± 12.7
Back of the head	- 0.9 ± 17.6	9.7 ± 11.9
Front of the neck	- 3.3 ± 19.6	2.1 ± 9.5
Front of the chest	- 24.5 ± 16.7	- 5.1 ± 10.3
Front of the waist	- 19.0 ± 14.9	- 6.2 ± 10.7
Sacrum	- 19.5 ± 19.3	47.8 ± 16.8
Right foot	0.8 ± 14.6	8.7 ± 12.2

* *Negative value denotes flexion.*

** *Negative value denotes radial deviation.*

Personal hygiene, food preparation and typical work functions required continuous wrist motion. Combing hair, perineal care, dental care, buttoning and unbuttoning, tying and untying ties/scarves and tying and untying laces required expression of wrist motion from 42° extension to 54° flexion and from 40° ulnar deviation to 15° radial deviation. Perineal care and the motions accompanying tying and untying scarves/ties required significantly greater amounts of flexion (54° and 51°, respectively) than did other aspects of personal care. Tying scarves and neckties also required the largest arc of radioulnar deviation, 55°, including both extreme radial (15°) and ulnar (40°) deviation (Figure 6).

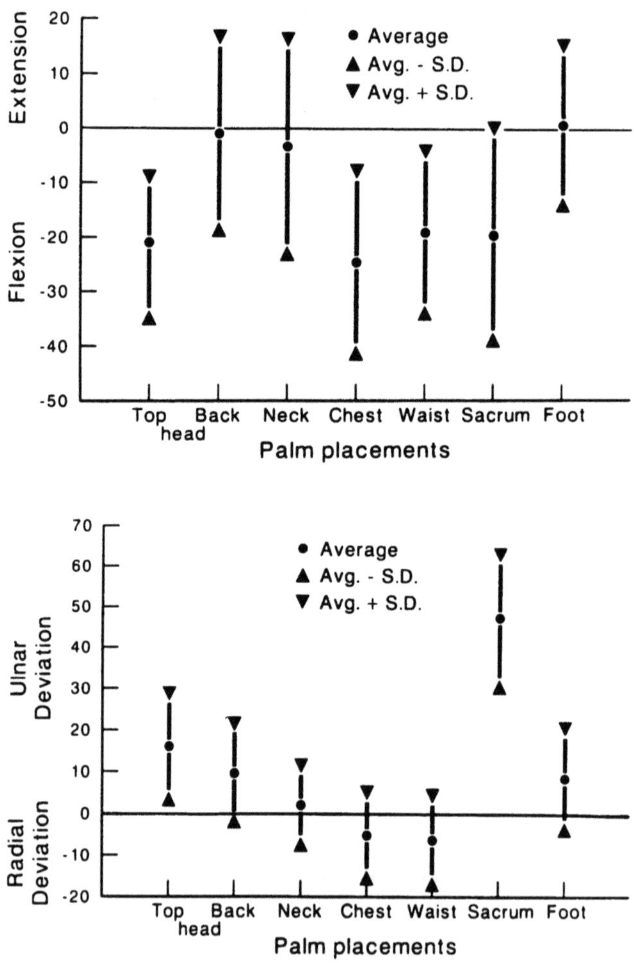

FIGURE 5. Category one: Seven activities requiring positioning of the wrist for personal care and hygiene (a) extension/flexion (top), (b) ulnar/radial deviation (bottom).

A more limited range of wrist motion was required for food preparation. Wrist motion from 42° extension to 37° flexion and 40° ulnar deviation to 12° radial deviation was required to perform the six activities in this category addressed. Turning a spatula (73° flexion/extension arc and 39° ulnar deviation) and opening the lid of a jar (47° radioulnar deviation) represented the extremes of motion presented by food preparation. Faucet and spatula use required the greatest extension (42° and 37°, respectively) (Figure 7).

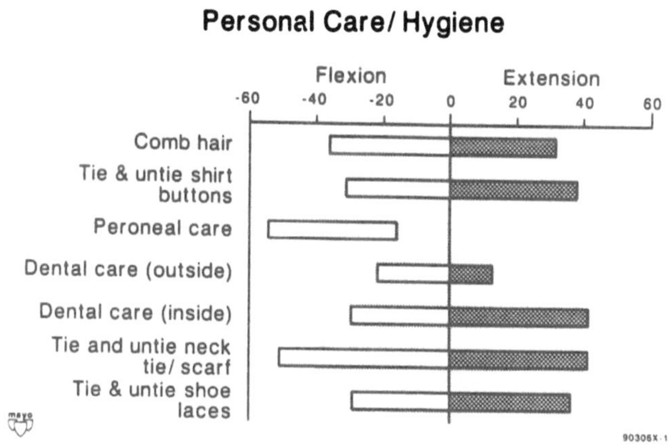

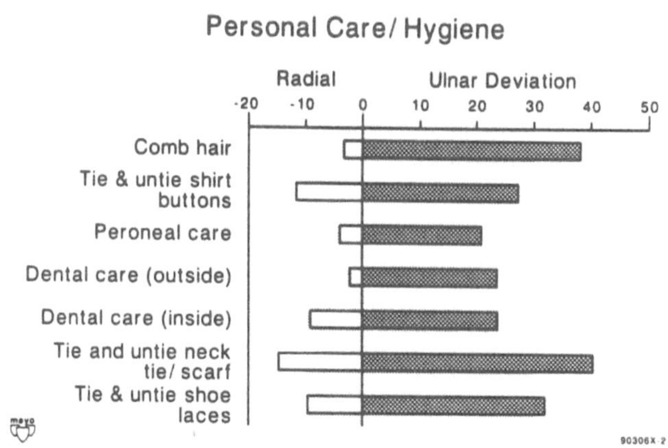

FIGURE 6. Range of wrist motion required during personal care/hygiene activities (a) extension/flexion, (b) ulnar/radial deviation. The grey striped area represents 70% of the maximum motion (40° extension, 40° flexion, 10° radial deviation and 30° ulnar deviation).

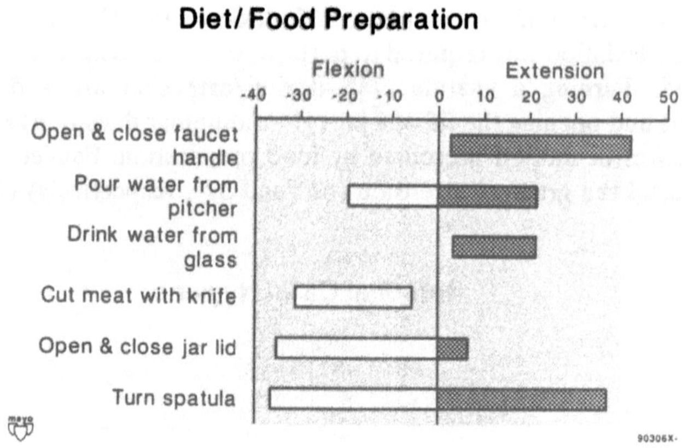

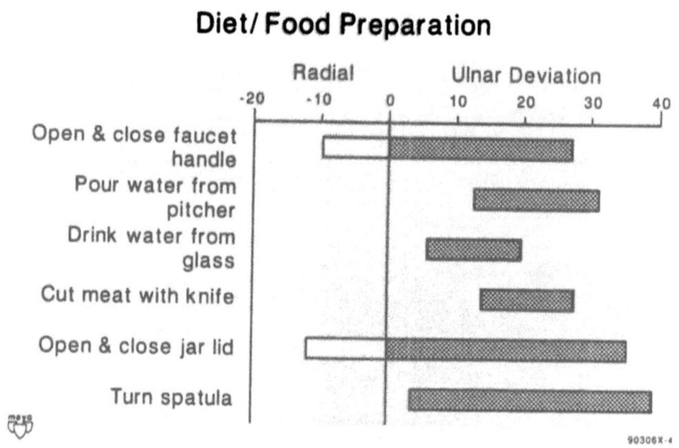

FIGURE 7. Range of wrist motion required during diet/food preparation activities (a) extension/flexion, (b) ulnar/radial deviation.

The category of typical work functions included manual labor activities associated with various forms of wrist motion (hammering, use of a screwdriver, turning doorknobs, keys and steering wheels, bringing a telephone to the ear, and rising from and sitting on a chair. The act of writing was included for comparison as it is an important task, though it demands little activity on the part of the wrist. Use of an arm rest for support when rising from a chair required the most extreme extension, 60°, and radial deviation, 17°, of all twenty-four (personal care, food preparation and work function combined) activities examined. Three different turning activities, hammering and phoning required extension slightly greater than 40°. Vertical insertion of screw while the subject was sitting required a large amount of flexion, 43°, while horizontal screw insertion was accomplished with ulnar deviation, 31°. Hand writing required the smallest motion arc, 14° extension/flexion and 12° radioulnar deviation (Figure 8).

To perform all twenty-four activities in a normal, comfortable, effective manner, a total of 60° wrist extension, 54° wrist flexion, 40° ulnar deviation and 17° radial deviation are essential. If total wrist motion was reduced to 40° each extension and flexion (70% of the maximum range of motion required), most, but not all of the studied activities could be performed by subjects. Thirteen of these were performed normally, and nine more could be performed adequately with complimentary forearm motion. Rising from a chair (60° extension) and perineal care (54° flexion) were performed satisfactorily with just 40° each flexion and extension, as splinting the wrists of our subjects in these positions demonstrated. In addition, patients splinted to allow only 30° radial and 10° ulnar deviation were able to perform fourteen of the studied activities normally, while ten other activities required only a few more degrees of radioulnar motion to be accomplished. From these studies it was determined that 40° extension, 40° flexion, and a combined 40° radioulnar deviation provided the minimal functional range of motion for "normal" populations (Ryu, et al., 1991).

Comparison of electro and hand-held goniometer findings in attempts to determine accuracy showed that the electrogoniometer recorded greater extension motion, but equal arcs of flexion/extension and radioulnar deviation than did the hand-held equipment. The difference in flexion/extension may be a reflection of the initial (zero) starting point (Table 2). Comparison of first and last electrogoniometer recordings of maximum motion proved the device to be extremely tight -- slippage was less than 1.2°. Reproducibility of electrogoniometer findings was demonstrated by repeat measurements of the same activity (Figure 4b). For each subject tested, standard deviations of three trials per task (Figure 4a) were 3° extension, 4° flexion, 3° ulnar and 3° radial deviation.

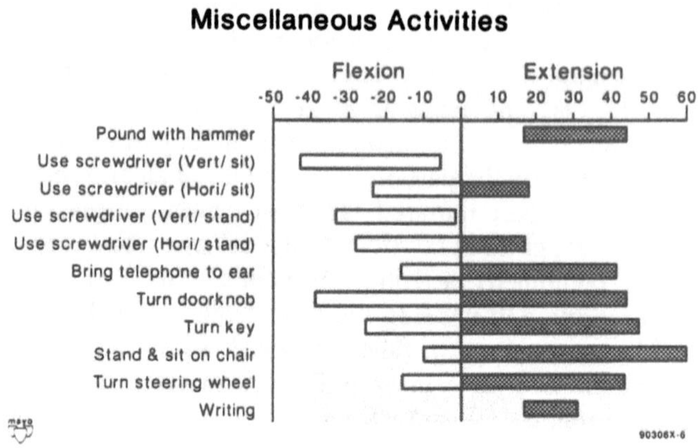

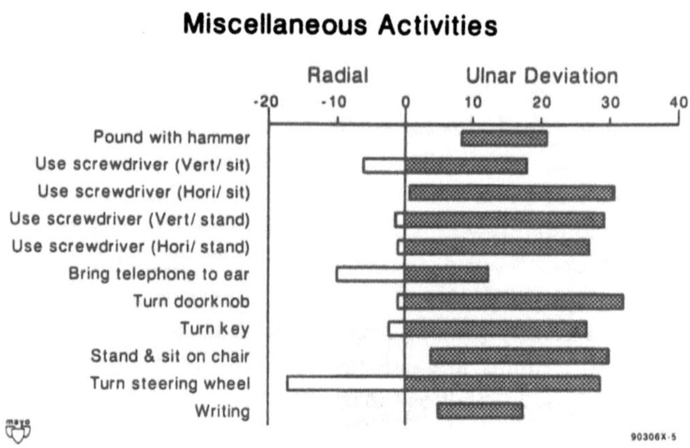

FIGURE 8. Range of wrist motion required during other frequently used ADLS (a) extension/flexion, (b) ulnar/radial deviation.

From the study by Palmer et al. (1985) it was found that the average maximum wrist arc of motion was 133° of flexion/extension and 40° of radial/ulnar deviation. When considering tasks related to personal hygiene, culinary skills, and other activities of daily living, an average range of 32° of flexion to 59° of extension was utilized with an average of 33° of total arc of motion (range 5° of flexion and 28° of extension). Radial deviation for these tasks averaged 9° and ulnar deviation averaged 12°. Specific tasks required specific ranges and directions of motion. For example, utilizing a knife requires minimal wrist motion. Many tasks required wrist motion which was centered relatively equally in all planes, such as buttoning a button, tying a shoe, stirring in a bowl, and opening and closing a faucet. Other tasks required motion predominately in the extension mode with variable amounts of radial/ulnar deviation, such as combing hair, wringing a washcloth, and eating with a fork. Additionally, some tasks required use of the so called "dart throwers motion" which incorporates a combination of extension and radial deviation to flexion and ulnar deviation, such as combing hair, wringing a washcloth, tying a shoe, and pouring from a pitcher.

Centroids (centers of motion) were defined for each task as the average location of the angular positions of the wrist for that specific task. The average centroid for personal hygiene was 13.8° of extension and 0.8° of ulnar deviation. The centroids for culinary skills averaged 11.5° of extension and 5.4° of ulnar deviation. In all, twenty-one of the twenty-four tasks had centroids in the extension mode and fifteen of the twenty-four tasks had centroids in the ulnar deviation mode. Carpentry tasks, as defined in Table 1, averaged 21° of flexion/extension and 17° of radial/ulnar deviation. Nearly all activities were performed in the extension mode, and some activities, such as using a power drill, required very little wrist motion. The centroid of motion for all tasks in the housekeeping classification were in the extension mode, and for the most part also in radial deviation mode. Tasks related to mechanics were carried out mostly in the extension mode with minimal radial/ulnar deviation. A minimal amount of wrist motion was necessary for the tasks defined in the secretarial section, where it was found that folding a letter and answering a telephone required predominately flexion mode. A wide variation in the ranges of motion in centroids of motion were utilized in the surgical task section, however, flexion was the most common mode for the given tasks. Finally, it was determined that based upon the evaluation of the fifty-two tasks outlined in Table 1, it appears that the functional range of motion of the wrist is between 5° of flexion, 30° of extension, 10° of radial deviation, and 15° of ulnar deviation. It was stressed that although operative procedures that limit wrist motion are not necessarily functionally detrimental, attempts should be made to place the wrist in a position such that this limited range of motion falls within the functional range as determined by this and other studies. From a clinical standpoint, future studies are needed to determine if surgical procedures on the wrist such as fusion, ligament reconstruction or joint prosthetics can accommodate such needs.

Other Goniometric Studies

Electrogoniometer measurements can also be used to adjunctively quantitate wrist motion requirements in patients with specific functional demands, such as athletic or artistic performances. Ryu et al. (1991) supplemented normal range of motion studies with evaluations of wrist movement required to perform basketball free-throws and piano exercises.

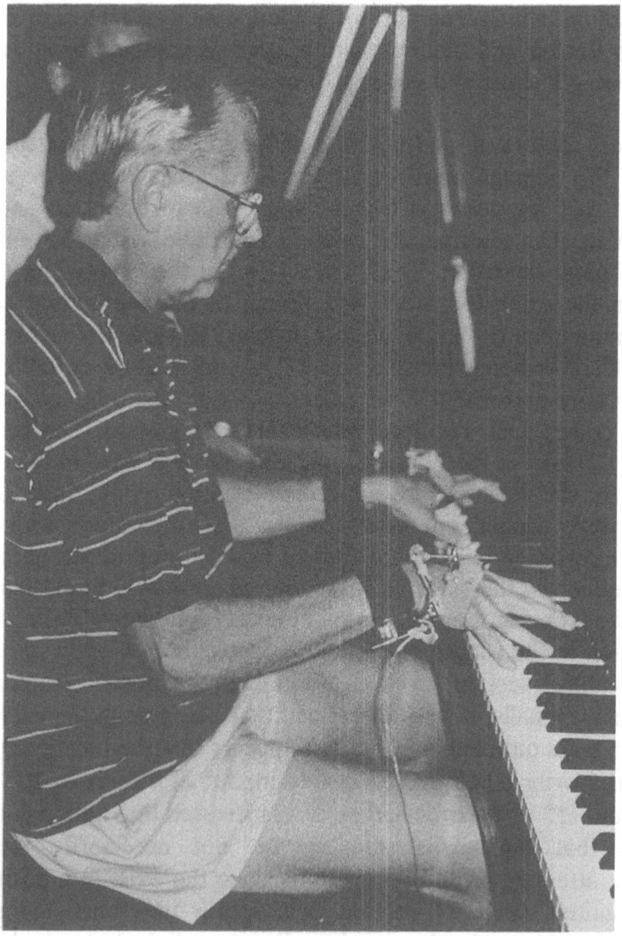

FIGURE 9. Wrist joint motion during piano playing.

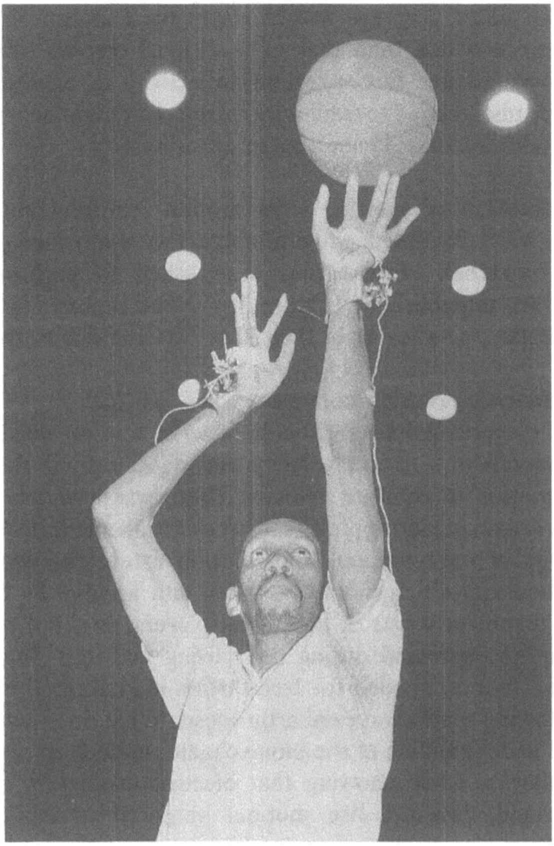

FIGURE 10. Wrist joint motion during basketball throwing.

In the piano studies (Figure 9), the wrist movement involved in standard warm-up exercises and classical examples of trills, arpeggios, octaves and broken octaves was examined (Chung, et al., 1991). In this study, a comparison was made between pianists trained under two different techniques (traditional and weight playing, a method in which the forearm is subject only to the force of gravity opposed by the hand's resistance to the keyboard -- forearm extensors serve as weak accessories attempting to prevent overuse syndrome.) Biaxial electrogoniometers were attached to the wrists of each player, and discrete fourier transform analysis was used to determine if differences existed between wrist motion required to play various types of music, or as expressed by the two groups.

Data analysis showed average range of motion (especially radioulnar deviation) during piano playing was greater than that of daily living activities.

Classical trills and arpeggios required more extensive wrist motion than did other musical pieces. Average weight player wrist motion in exercises and classical music was less than that of traditional pianists, although weight players showed greater flexion/extension activity in arpeggios and trills. Traditional pianists tended to exhibit increased radioulnar activity. This serves as a valid model for wrist kinematics in musicians.

In consideration of sporting wrist motion requirements, a study was developed in which biaxial electrogoniometers were attached to the wrists of six NACC basketball players who were asked to perform free-throws (Ohnishi, et al., unpublished) (Figure 10). None reported a hampering of their wrist motion, or altering of their shot patterns due to the devices.

Motion analysis revealed consistency and particularity of wrist motion during each player's shooting. Dominant wrist motion divided into three phases (acceleration, constant velocity and deceleration) showed uniform ulnar wrist motion at constant velocity. There was a variety in radioulnar wrist motion direction among players in acceleration and deceleration phases. A great range of wrist motion (120° flexion/extension) was required for basketball shooting, which should be taken into account by hand surgeons treating the dominant wrists of basketball players, as this far surpasses the range of motion expressed during daily living activities. Interestingly, the range of wrist motion needed for free-throws in basketball is close to that reported following scaphotrapezial arthrodesis. What this study provides is a fundamental understanding of the biomechanics underlying wrist function in basketball players, while showing that electrogoniometers can accurately assess the rapid, precision-fire motions required in athletic endeavors (Figure 7).

In conclusion, the electrogoniometer seems to provide a reliable and reproducible tool for the evaluation of functional wrist motion. The insight it provides into the functional assessment of wrist pathology and potential treatment outcomes offers invaluable help in the selection of therapeutic techniques to aid patients suffering wrist impairment. If this selection can be tailored to meet occupational and avocational wrist motion requirements, hand surgeons will be that much further toward understanding the wrist and its potential for intricate motion, and returning patients to the opportunities their dexterity has led them to expect.

References

American Academy of Orthopaedic Surgeons. Joint Motion: Method of Measuring and Recording. Chicago, The American Academy of Orthopaedic Surgeons, 1965.

An KN, Chao EYS: Kinematic Analysis of Human Movement. Ann Biomedical Engrg 1984;12:585-597.

Ayoub MA, Ayoub MM, Ramsey JD: A Stereometric System for Measuring Human Carpal Motion. Hum Factors 1970;12:523-535.

Boone DC, Azen SP: Normal Range of Motion in Joints of Male Subjects. J Bone and Joint Surg 1979;61-A:756-759.

Brumfield R, Nickel V, Nickel E: Joint Motion in Wrist Flexion and Extension. South Med J 1966;59:909-910.

Chao EYS: Justification of Triaxial Goniometer for the Measurement of Joint Rotation. J Biomech 1980;13:989-1006.

Chao EYS: Experimental Methods for Biomechanical Measurements of Joint Kinematics. In: CRC Handbook of Engineering in Medicine and Biology, West Palm Beach, Florida, CRC Press, 1978:385-411.

Chao EYS, Morrey BF: Three-dimensional Rotation of the Elbow. J Biomech 1978;11:57-73.

Chung IS, Ryu J, Ohnishi N, et al: Wrist Motion Analysis in Pianists. Submitted to the J Med Probs Performing Artists, April 1991.

Cooney WP, Beckenbaugh RD, Linscheid RL: Total Wrist Arthroplasty. Clin Orthop 1984;187:121-128.

DeRoos JP, Chao EY, Cooney WP: A Biomechanical Method for Evaluation of Normal and Pathologic Thumb Function. Proc 23rd Ann Orth Res Soc 1977;2:201.

Eberhart HD, Inman VT: An Evaluation of Experimental Procedures used in a Fundamental Study of Human Locomotion. Ann NY Acad Sci 1967;51:1213-1228.

Engen TJ, Spencer WA: Method of Kinematic Study of Normal Upper Extremity Movement. Arch Phys Med Rehabil 1968;49:9-12.

Erdman AG, Dorman R, Isaacson R, Scott D, Weinberg I: Stereoscopic Instrumentation for Analyzing Gross Motion. ASME Paper 1976; No.76-DET-23.

Hayes WC, Feldman JM, Oatis C, Nixon JE: Gait Analysis by Multiaxial Accelerometry. Trans. 24th Orth Res Soc 1978;3:104.

Karpovich PV, Karpovich GP: Electrogoniometer: A New Device for Study of Joints in Action. Fed Proc 1959;18:79.

Kleinman WB, Steichen JB, Strickland JW: Management of Chronic Rotary Subluxation of the Scaphoid by Scapho-Trapezio-Trapezoid Arthrodesis. J Hand Surg 1982;7:125-136.

Kraft G, Detels P: Position of Function of the Wrist. Arch Phys Med Rehabil 1972; 53:272-275.

Morrey BF, Askew LJ, An KN, Chao EYS: A Biomechanical Study of Normal Functional Elbow Motion. J Bone Joint Surg 1981;63-A:87f2-877.

Morris JRW: Accelerometry – A Technique for the Measurement of Human Body Movement. J Biomech 1973;6:729-736.

Ohnishi N, Ryu J, Rowen B, Chung IS: Wrist Motion Analysis in Basketball Shooting. Material yet to be published.

Palmer AK, Werner FW, Murphy D, Glisson R: Functional Wrist Motion: A Biomechanical Study. J Hand Surg 1985;10-A(1):39-46.

Palmer AK, Dobyns JH, Lincheid RL: Management of Post Traumatic Instability of the Wrist Secondary to Ligament Rupture. J Hand Surg 1978;3:507-532.

Pryce JC: The Wrist Position between Neutral and Ulnar Deviation that Facilitates the Maximum Power Grip Strength. J Biomech 1980;13:505-511.

Ryu J, Cooney WP, Askew LJ, An KN, Chao EYS: Functional Ranges of Motion of the Wrist Joint. J Hand Surg 1991; 16A:409-419.
Sarrofian S, Melamed JL, Gosgarian GM: Study of Wrist Motion in Flexion and Extension. Clin Orthop 1977;126:153-159.
Speltz S, Schutt A, Beckenbaugh R: Functional Wrist Position for Arthrodesis. J Hand Surg 1983;8:627.
Volz RG, Lieb M, Benjamin J: Biomechanics of the Wrist. Clin Orthop 1980;149:112-117.
Watson HK, Ryu J, DiBella A: An Approach to Kienböcks Disease. J Hand Surg 1985:10-A:179-187.
Watson HK, Hempton RF: Limited Wrist Arthrodesis. J Hand Surg 1980;5(4):320-327.
Yoon YS: Analytical Development in Investigation of Wrist Kinematics. J Biomech 1979;12:613-621.
Youm Y, Flatt AE: Kinematics of the Wrist. Clin Orthop 1980;149:21-32.
Youm Y, McMurtry RY, Flatt AE, Gillespie TE: Kinematics of the Wrist. J Bone and Joint Surg 1978;60-A(4):423-431.

Chapter 4

Individual Carpal Bone Motion

M. Garcia-Elias, E. Horii, and R.A. Berger

Introduction

Wrist motion is the result of an interaction and accumulation of carpal kinematics occurring at the different levels of this complex and composite joint (Kapandji 1987; Kauer 1974; Lange, et al., 1985; Lange 1987; Linscheid 1986; Ruby, et al., 1988; Talesnik 1985). Understanding the individual carpal bone kinematics is of value not only from a basic perspective, but also from a clinical point of view, for many problems arising in the wrist are in fact the result of an alteration of the intracarpal motion (Berger, et al., 1982; Garcia-Elias, et al., 1989; Horii, et al., 1991; Ruby, et al., 1987; Smith, et al., 1989a.; Smith, et al., 1989b.). Its knowledge, therefore, is essential for a proper diagnosis and surgical treatment. Understanding carpal kinematics is also important to evaluate functional performance of the wrist under normal and pathological conditions, not to mention for the design of prosthetic devices (Chao and An, 1982; Youm, et al., 1978; Youm 1979).

The first published account of normal carpal kinematics is attributed to Bryce who, in 1896, interpreted planar radiographs of his own wrist (Bryce 1896). Remarkably, this was just one year following Roentgen's discovery of X-rays. Since then, several other approaches have been utilized to define individual carpal bone motion: direct visualization of dissected specimens (Johnston 1907), uniplanar radiographs (Von Bonin 1919), measurement of displacements of implanted wires into the carpal bones (Cyriax 1926), fluoroscopy (Wright 1935), cineradiography (Arkless 1966; Kauer 1974), light emitting diodes attached to the moving carpal bones (Berger, et al., 1982).

To avoid the effects of protruding wires or transducers, Youm (Andrews and Youm, 1979; Youm, et al., 1978; Youm 1979) implanted fine metal markers into selected carpal bones and using both biplanar radiography and cineradiography described the principle according to which the carpus keeps a constant "height ratio" during radioulnar deviation; besides, both flexion-extension and radioulnar deviation centers of rotation were defined and located. Their methods, however, did not allow quantification of the relative intracarpal motion.

Berger et al. (1982) were the first to report a quantitative analysis of the relative motion between the carpal bones. By implanting sound sources in selected carpal bones, and using a three-dimensional sonic digitizer, they recorded screw displacement motion for each carpal bone relative to an X,Y,Z coordinate system based on the distal radius. More recently Lange, using a stereometric measurement system, also studied the kinematic behavior of the normal carpal bones (Lange, et al., 1985; Lange 1987).

In the same category as the stereometric method, the biplanar radiographic method of kinematic analysis has long been used in this field. Although initially introduced by Chao and Morrey (1978), to study elbow motion, this method has been utilized, and proven useful, in many other areas. In the wrist area, several studies addressing different aspects of the kinematics have been performed at the Mayo Clinic Biomechanics Laboratory (Garcia-Elias, et al., 1989; Horii, et al., 1991; Ruby, et al., 1987; Ruby, et al., 1988; Smith, et al., 1989a.; Smith, et al., 1989b.).

In this chapter a short description of the methodology utilized will be presented, followed by a brief review of the results obtained in different conditions of carpal integrity.

Methodology of Intracarpal Kinematic Analysis

Radiologically normal fresh frozen human cadaveric wrists have been utilized in these experiments. The first step in the process involves marker insertion into each carpal bone (Figure 1A-B). Through a longitudinal dorso-radial incision the dorsal capsule is exposed. The extensor retinaculum is identified and retracted proximally. The transverse dorsal intercarpal ligament is localized and preserved. The distal carpal row bones, except the trapezium, are exposed distal to the transverse intercarpal ligament, while the distal radius, proximal pole of the scaphoid, the lunate and the triquetrum are exposed through small arthrotomies by flexing the wrist. A second dorsoradial skin incision is placed at the base of the thumb to insert markers into the trapezium and distal part of the scaphoid. Care is taken to avoid injuring any intrinsic or extrinsic ligament during marker insertion. Four or five small drill holes are made in each bone. Different sizes of radio dense tantalum spheres are inserted into the predrilled holes. For a complete kinematic analysis, at least three noncollinear markers must be implanted. To increase the accuracy and to be able to exclude occasionally loose markers, four or five well-separated spheres are currently inserted. The position of the markers is secured by a small amount of cyanoacrylate, after which all incisions are surgically closed.

The second step in the process involves securing each specimen to a testing frame and loading the wrist muscles. The five main muscles involved in wrist motion - Extensor Carpi Ulnaris (ECU), Extensor Carpi Radialis Longus and Brevis (ECRL and ECRB), Flexor Carpi Radialis (FCR) and Flexor Carpi Ulnaris (FCU)- are exposed proximally in the forearm, and Nº2 Mersilene sutures are placed for tendon loading. Each specimen is rigidly mounted in a specially designed Plexiglas frame by means of two or three Steinmann pins drilled through the radius and the ulna with the forearm in neutral axial rotation. The wrist motor tendons are connected by heavy fishing line to calibrated springs, controlled by a turnscrew. In this way, the wrist specimen can be balanced and placed into any wrist position, by adjusting the individual spring tensions to simulate the physiologic tendon tension. The applied load to each tendon in these experiments is as shown in Table 1. These tension values are based on a number of studies of the physiologic cross-sectional areas, and relative electromyographic activity of each tendon in each wrist position (Smith, et al., 1989b.).

TABLE 1. *Tendon-loaded wrist positioning: applied tension (N)*

	ECRB	*ECRL*	*ECU*	*FCR*	*FCU*	*Total*
Neutral	6.5	6.5	6.5	6.5	7.7	33.7
Extension	8.8	6.7	10.7	—	—	26.2
Flexion	—	—	—	9.4	8.7	18.1
Radial deviation	9.2	9.7	—	9.1	—	28.0
Ulnar deviation	—	—	10.8	—	9.8	20.6

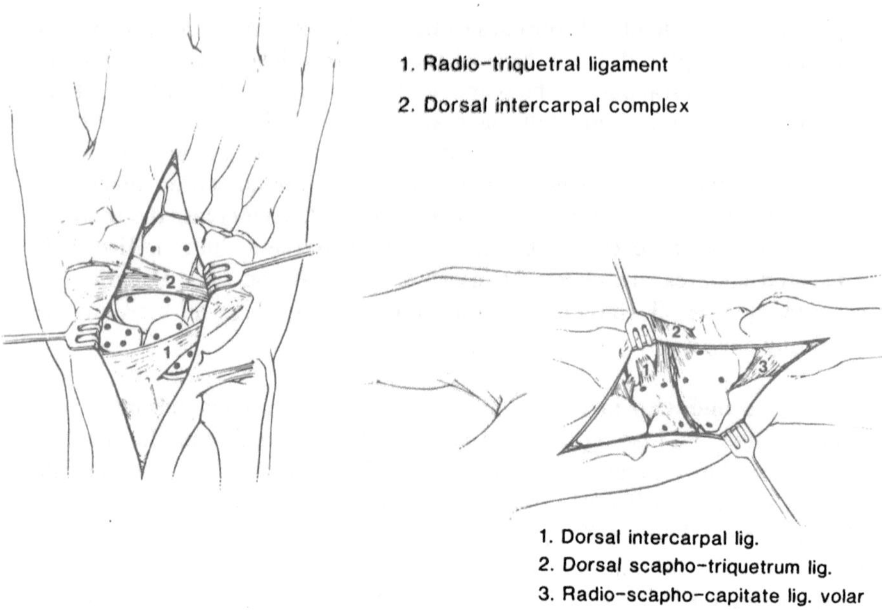

FIGURE 1A-B: Schematic representation of exposure of the dorsal ligaments and marker insertion into the relevant carpal bones. A) Longitudinal mid-dorsal incision. B) Dorso-radial incision (reprinted with permission of Smith, et al., J Orthop Res 1989;7:590-598).

The wrist specimen and frame is then placed within a biplanar radiographic apparatus. The cassette holder maintains the two x-rays in an exact orthogonal relationship to two fixed x-ray tubes (Figure 2). The cassette holder contains a system of reference markers the position of which are predetermined by a calibration procedure. The closest markers to the films are used to create a reference coordinate axis system. The markers in the wall nearest to the roentgen tubes are used to calculate the position of the x-ray sources with respect to the reference frame.

Biplanar radiographs are obtained in five wrist positions (neutral, flexion, extension, radial deviation and ulnar deviation) with the wrist in the loaded condition. The images of the markers in the biplanar x-rays are numbered according to a standardized pattern. The two-dimensional position of each point is measured using a digitizer and stored in a PC-computer. The data is then analyzed utilizing a especially designed software which includes four different programs:

- The first program (called DIGIT) coordinates the systematic digitization of both carpal and reference markers;

- The second (called CALC), Calculates the three-dimentional location of each marker with respect to the global coordinate system;

- A third program (called RIG) determines the absolute motion of the selected bones with respect to the global coordinate system, based on a theoretical method reported by Spoor and Velpaus;

- The fourth program (called REL) calculates the relative motion of the selected bones with respect to the radius or whichever reference is chosen.

The relative motion between each rigid body pair is described as a rotation around, and translation along a unique axis, following the "screw displacement axis" concept (Chao and An, 1982). Orientation and position of every screw axis is described on an XYZ axis system, such that X lies along the radius, Y is transverse across the wrist, and Z is in the sagittal plane perpendicular to the other two axes. Motion about the X axis is, therefore, pronation-supination, motion about the Y axis is flexion-extension, and motion about the Z axis is radial-ulnar deviation (Figure 3).

The accuracy of this technique was validated in a previous investigation by measuring the orientation and separation of phantom markers embedded in a plexiglas cube (Chao and Morrey, 1978). This kinematic analysis system has shown an accuracy to within 0.4 mm displacement and 2° angular rotation.

Normal Carpal Kinematics

None of the muscles acting on the wrist are inserted on the proximal carpal row. They are attached distally, either to the base of the metacarpals (FCR, ECRL, ECRB, ECU) or to the bones of the distal carpal row (FCU). The moments created as a result of the contraction of these muscles will generate motion starting at the distal carpal row (Ruby, et al., 1988). Table 2 summarizes some of the results of the kinematic analysis using five fresh human cadaver wrists performed by Horii et al. (1991).

In normal wrists, very little intracarpal motion exist between the bones of the distal carpal row (Berger, et al., 1982; Lange, et al., 1985; Ruby, et al., 1988). From full flexion to extension, no more than 9° angular rotation at the hamate-capitate joint, no more than 6° at the capitate-trapezoid joint, or more than 12° at the trapezoid-trapezium joint, were recorded by Ruby et al. (1988). The bones of the distal carpal row can be thought of as one functional unit. In flexion of the wrist, they all follow a rotation about an axis which obviously implies flexion, but also an ulnar deviation. In extension the tendency of all distal carpal bones is to go into a slight radial deviation (Horii, et al., 1991; Ruby, et al., 1988). The palmar concavity of the carpus changes little in flexion-extension of the wrist, for only a very small amount of pronation-supination motion between the bones of the distal carpal row is allowed by the stout interosseous ligaments connecting them.

FIGURE 2: Each specimen is mounted within a Plexiglas frame and biplanar X-rays obtained. The radiographic cassette holder maintains the two radiographic cassettes in an exact orthogonal relationship and has incorporated radiographic markers for the creation of a reference coordinate system within which carpal bone motion can be described (reprinted with permission of Smith, et al., J Orthop Res 1989;7:590-598).

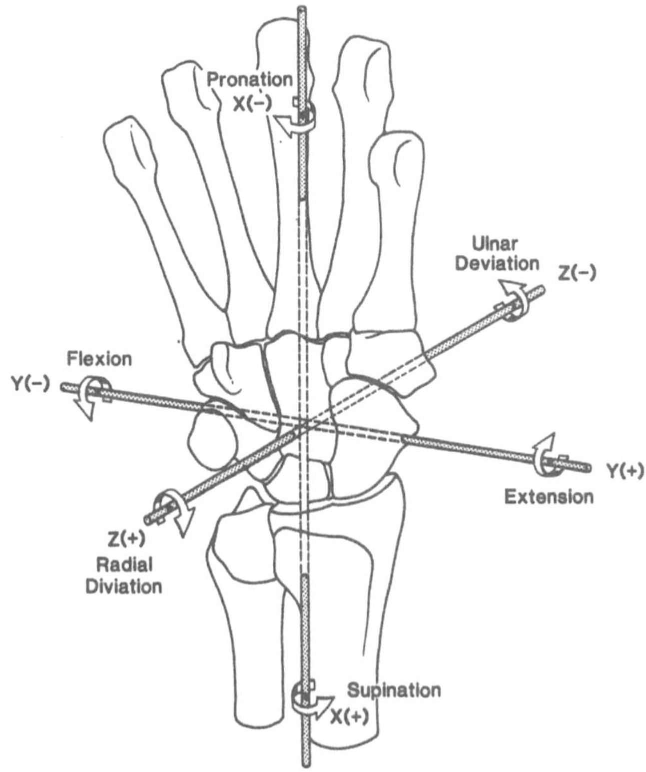

FIGURE 3: XYZ coordinate system used to describe screw axis orientation and position, viewing the carpus from an antero-lateral position (reprinted with permission of Garcia-Elias, et al., J Hand Surg 1989;14A:791-799).

Acting as a unit, the bones of the distal carpal row move synergistically in lateral deviations of the wrist. In radial deviation of the wrist, the distal carpal row bones extend, supinate and radial deviate. In ulnar deviation, they flex, ulnarly deviate and pronate (Horii, et al., 1991; Kauer 1974; Lange, et al., 1985; Smith, et al., 1989b.).

The bones of the proximal carpal row appear to be less tightly bound to one another than the bones of the distal carpal row (Berger, et al., 1982; Horii, et al., 1991; Kauer 1974; Lange 1987). Despite differences in angular rotation, however, all proximal carpal row bones move in approximately the same direction no matter what global wrist motion occurs (Rudy, et al., 1988). Thus, it is reasonable to consider the proximal carpal row as another functional unit, acting as an intercalated segment between the distal row and the radius (Kauer 1974; Linscheid 1986).

TABLE 2. Normal carpal bone motions in flexion, extension and deviations of the hand, with respect to the radius, expressed using the "screw axis displacement" concept as found by Horii et al. (1991). (Average values of five consecutive specimens.)

1: From neutral to extension

	Orientation of screw axis (*)			Rotation	
Moving bone	X	Y	Z	(mean	SD)
Scaphoid	0.10	0.98	0.05	56.1	9.2
Lunate	0.09	0.98	0.15	31.2	4.9
Triquetrum	0.12	0.98	0.05	41.6	5.1
Capitate	0.00	0.99	0.03	64.2	6.8
Hamate	-0.01	0.99	0.00	65.0	5.7

2: From neutral to flexion

	Orientation of Screw Axis (*)			Rotation	
Moving bone	X	Y	Z	(mean	SD)
Scaphoid	-0.04	-0.96	-0.23	55.4	16.9
Lunate	0.01	-0.90	-0.35	45.1	19.8
Triquetrum	-0.01	-0.92	-0.30	47.7	21.2
Capitate	0.02	-0.99	-0.10	77.0	18.7
Hamate	0.04	-0.99	-0.10	72.2	17.0

3: From neutral to radial deviation

	Orientation of Screw Axis (*)			Rotation	
Moving bone	X	Y	Z	(mean	SD)
Scaphoid	0.24	-0.72	0.62	12.8	6.9
Lunate	0.14	-0.78	0.63	13.0	6.9
Triquetrum	0.16	-0.49	0.83	11.8	7.2
Capitate	0.10	0.06	0.97	24.2	8.3
Hamate	0.15	0.15	0.97	23.9	7.3

4: From neutral to ulnar deviation

	Orientation of Screw Axis (*)			Rotation	
Moving bone	X	Y	Z	(mean	SD)
Scaphoid	-0.24	0.74	-0.55	22.7	6.0
Lunate	-0.13	0.80	-0.54	25.4	7.3
Triquetrum	0.01	0.64	-0.84	23.3	7.3
Capitate	-0.21	-0.15	-0.95	28.7	8.8
Hamate	-0.16	-0.08	-0.96	28.0	9.7

(*) X-axis: + supination - pronation
Y-axis: + extension - flexion
Z-axis: + radial dev. - ulnar dev.

According to Ruby et al.'s experiments (1987; 1988), from full flexion to full extension of the wrist, the scaphoid rotates an average of 80.3° with respect to the radius, while the lunate rotates 58.6° and the triquetrum rotates 70.9°. Horii et al. (1991) found different relative values for the scaphoid, lunate and triquetrum rotations from global wrist flexion to extension (110°, 76°, and 88° respectively (Table 2). According to Ruby et al. (1988), the average scapholunate and lunotriquetral intracarpal rotations are 24° and 18° respectively. According to Horii's data, however, there is almost three times as much motion at the scapholunate interval than at the lunotriquetral joint (34° and 12°, respectively). Also Smith et al. (1989a.; 1989b.) and Lange et al. (1985; 1987) found a scapholunate relative motion larger than 30°. Despite differences in absolute values of angular rotation, however, the direction of motion found by all the authors is quite consistent: during wrist flexion, scaphoid, lunate and triquetrum go into flexion and ulnar deviation, while during wrist extension they extend and radial deviate (Berger, et al., 1982; Garcia-Elias, et al., 1989; Horii, et al., 1991; Lange 1987; Ruby, et al., 1988; Smith, et al., 1989b.).

The contribution of the midcarpal joint to normal wrist flexion-extension is also an interesting issue. If only the central part of the carpus (the capitate-lunate-radius linkage) is considered, the radiocarpal and midcarpal motion contribute equally to total ulnar motion (Garcia-Elias, et al., 1989; Linscheid 1986). However, if motion is recorded on the lateral part (the radius-scaphoid-trapezium linkage), more than two-thirds of the global arc of movement occur at the radioscaphoid interval (Garcia-Elias, et al., 1989; Ruby, et al., 1988; Smith, et al., 1989b.). According to that, an arthrodesis of the lunocapitate joint theoretically would result in a larger restriction of the global wrist motion than a fusion of the scaphotrapezial-trapezoidal joint.

According to Ruby et al. (1988), from full radial deviation to full ulnar deviation, the scaphoid rotates an average of 51°, while the lunate rotates 35°, and the triquetrum rotates 28°. Horii et al. (1991), by contrast, could not find such differences between the three bones (average 36°, 38°, and 35° for the scaphoid, lunate and triquetrum, respectively). In both studies, however, this rotation was found to occur about an axis implying radial deviation but also extension. Despite controversies about the amount of rotation experienced by each individual carpal bone during this motion, all investigators agree in that the three proximal carpal bones move synergistically from a flexed position in radial deviation to an extended position in ulnar deviation, (Berger, et al., 1982; Garcia-Elias, et al., 1989; Lange, et al., 1985; Smith, et al., 1989a.; Smith, et al., 1989b.; Weber 1984). This flexion-extension adaptive mechanism, present in normal wrists, allows a constant spatial congruency between the distal carpal row and the radius no matter what wrist position is adopted (Kapandji 1987; Kauer 1974; Linscheid 1986).

Pathokinematics of the Wrist

The above mentioned carpal shifting mechanism by which the proximal carpal row is able to remain stable while adapting its position to the continuously changing space between the distal carpal row and the two forearm bones is one of the most important features of the wrist (Garcia-Elias, et al., 1989; Kapandji 1987; Kauer 1974; Linscheid 1986). The loss of this mechanism usually results in articular incongruency, and therefore, cartilage degeneration and progressive adjacent ligament disruption, which in its turn creates more instability and abnormal kinematics. At the Biomechanics laboratory of Mayo Clinic four different conditions known for their ability of altering carpal kinematics have been studied. What follows is a brief review of the results obtained.

Scapholunate Dissociation

Individual scaphoid and lunate motion before and after sectioning of different parts of the scapholunate ligamentous complex was studied using four cadaver specimens (Ruby, et al., 1987). Disruption of the dorsal portion of the scapholunate interosseous ligament causes a significant change in the normal relationship between the scaphoid and the lunate. In all four specimens, the scaphoid was observed to collapse primarily in flexion and pronation with respect to the radius. By contrast, the lunate position changed very little: only an average of 4.9° towards extension. The presence of the radioscapholunate ligament was found not to prevent neither a dorsal subluxation of the scaphoid, nor an average 3 mm scaphoid-to-lunate gap formation, if disruption of both the palmar and dorsal portions of the scapholunate interosseous ligament was complete.

Limited Intercarpal Arthrodeses

Limited intercarpal fusions, especially those crossing the midcarpal joint, were found to disturb intracarpal kinematics (Garcia-Elias, et al., 1989). In one of the experiments, the relative kinematic behavior of selected carpal bones, before and after simulation of two different fusions, was analyzed. Fusions were simulated by means of two non-parallel Herbert screws inserted as shown on Figure 4. Both the STT and SC fusions produced only a modest decrease in wrist global range of motion. However, they alter very significantly intracarpal kinematics. In ulnar deviation, for instance, the scaphoid of normal wrists rotates around an screw axis following extension, ulnar deviation and pronation. After simulating an STT fusion, the flexion-extension component of scaphoid motion decreased significantly. The resultant screw axis after an STT fusion was observed to have a direction somewhat parallel to that found for the capitate (Figure 5). In other words, the scaphoid, after these procedures, was found to act as if it was another bone of the distal row. Consequently, because the scaphoid is unable to palmar flex when the trapezium approaches the radius, compressive stress in the radioscaphoid joint is likely to increase in radial deviation, while greater tensile stress in the scapholunate ligaments will result in ulnar deviation because the scaphoid cannot extend (Figure 6).

Scaphoid Waist Unstable Fractures

Utilizing the same quantitative method of carpal kinematic analysis, the kinematic effects of a scaphoid waist osteotomy were also studied (Smith, et al., 1989a.; Smith, et al., 1989b.). The testing of five osteotomized specimens revealed that the two scaphoid fragments not only move independently, but their relative motion is more complex and significant than previously recognized. After scaphoid osteotomy, the proximal scaphoid motion during extension increased from an average of 29° to 49°, while the distal scaphoid had less rotation from 29° to 23°. That implies a dorsal and radial angulation at the fracture site. This flexion collapse is likely to result in the "humpback" deformity that is commonly seen in chronic scaphoid nonunions.

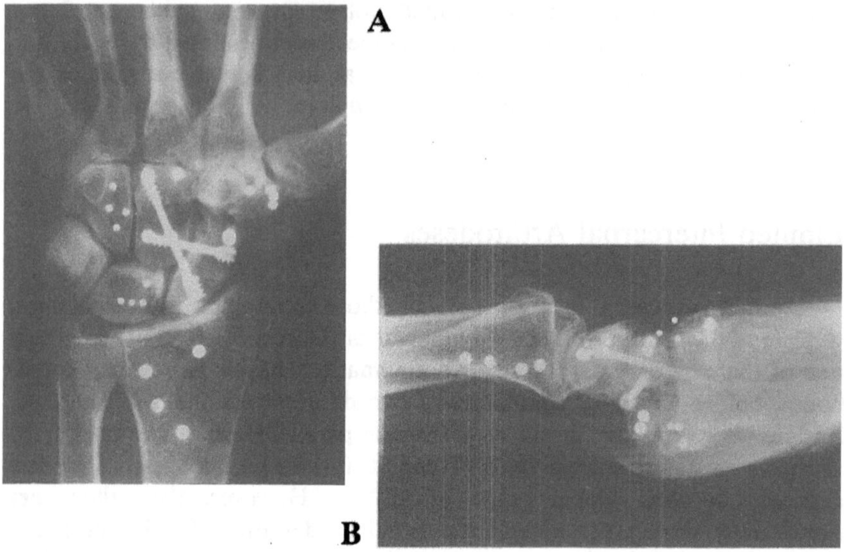

FIGURE 4: Biplanar radiographs demonstrate markers embedded into each carpal bone and orientation of the Herbert screws used to simulate an SC fusion in AP view (A) and lateral view (B) (reprinted with permission of Garcia-Elias, et al., J Hand Surg 1989;14A:791-799).

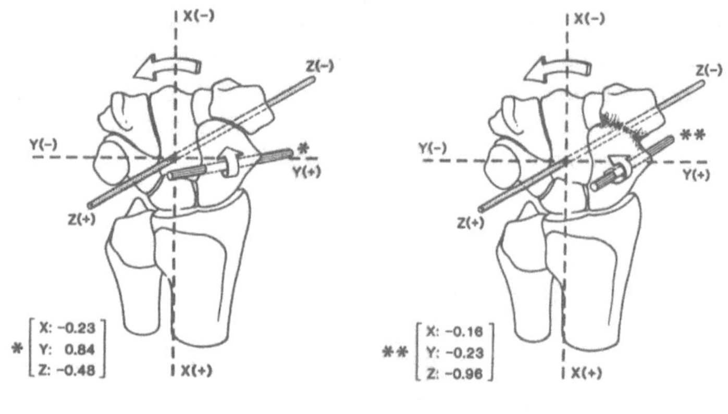

FIGURE 5: During ulnar deviation the scaphoid of an intact wrist rotates about a screw axis (*) allowing extension (Y+), ulnar deviation (Z-), and pronation (X-). After an STT fusion, the scaphoid kinematics changes significantly. The screw axis in this new situation (**) is almost parallel to the one found for the capitate (shaded cylinder). That axis basically implies ulnar deviation (Z-) and a small amount of flexion (Y-) and pronation (X-), which is the typical behavior of the distal carpal row bones during ulnar deviation (reprinted with permission of Garcia-Elias, et al., J Hand Surg 1989;14A:791-799).

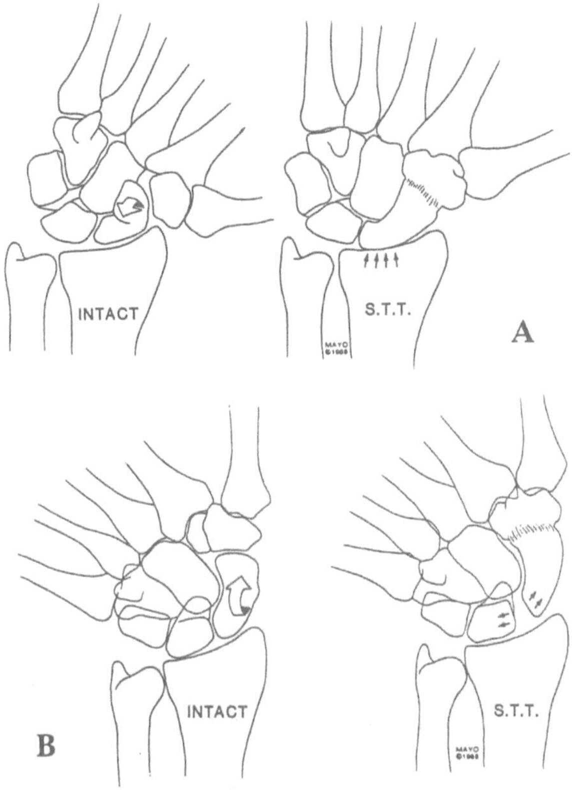

FIGURE 6: Since the scaphoid behaves as another distal carpal row bone after an STT fusion, increasing pressure at the radioscaphoid joint is likely to appear during radial deviation (A), while increasing tension on the scapholunate ligaments during ulnar deviation (B) is to be expected (reprinted with permission of Garcia-Elias, et al., J Hand Surg 1989;14A:791-799).

Lunotriquetral Dissociations

Another experiment using the above described biplanar method of kinematic analysis investigated the effects of sectioning the ligaments supporting the lunotriquetral (LTq) joint in five cadaver specimens (Horii, et al., 1991). The ligaments were sectioned in two stages. In stage I, a complete sectioning of both the dorsal and palmar LTq ligaments and the interosseous membrane was performed. In stage II, further sectioning of both the dorsal radiotriquetral and dorsal scaphotriquetral ligaments was performed. After both stage I and II, all the intercarpal joints (and especially the LTq joint) exhibited altered kinematics, with the lunate adopting a palmar-flexed position, and the triquetrum rotating into supination and eventually VISI pattern of carpal instability.

References

Andrews JG, Youm Y: A biomechanical investigation of wrist kinematics. J Biomechanics 1979;12:83-93.

Arkless R: Cineradiography in normal and abnormal wrists. Am J Roentgenology 1966;96:837-844.

Berger RA, Crowninshield RD, Flatt AE: The three-dimensional rotational behaviors of the carpal bones. Clin Orthop 1982;167:303-310.

Bryce TC: On certain points in the anatomy and mechanism of the wrist joint reviewed in the light of a series of roentgen ray photographs of the living hand. J Anat Physiol 1896;31:59-79.

Chao EYS, An KN: Perspectives in measurements and modeling of musculoskeletal joint dynamics. In Huiskes R, Van Campen D, Wijn J (eds.): Biomechanics: Principles and Applications. The Hague, Martinus Nijhoff Publishers, 1982:pp1-18.

Chao EYS, Morrey BF: Three-dimensional rotation of the elbow. J Biomechanics 1978;11:57-73.

Cyriax EF: On the rotatory movements of the wrist. J Anat 1926;60:199-201.

Erdman AG, Mayfield JK, Dorman F, Wallrich M, Dahlof W: Kinematic and kinetic analysis of the human wrist by stereoscopic instrumentation. J Biomechanical Engineering 1979;101:124-133.

Garcia-Elias M, Cooney WP, An KN, Linscheid RL, Chao EYS: Wrist kinematics after limited intercarpal arthrodesis. J Hand Surg 1989;14A:791-799.

Horii E, Garcia-Elias M, An KN, Linscheid RL, Bishop AT, Cooney WP, Chao EYS: A Kinematic Study of Lunotriquetral dissociations. J Hand Surg 1991;16(2): 355-362.

Johnston HM: Varying positions of the carpal bones in the different movements at the wrist. Part I: J Anat Physiol 1907;41:109-122. Part II: J Anat Physiol 1907;41:280-292.

Kapandji A: Biomécanique du carpe et du poignet. Ann Chir Main 1987;6:147-169.

Kauer JMG: The interdependence of carpal articulation chains. Acta Anat 1974;88:481-501.

Lange A de: A kinematic study of the human wrist joint. Doctoral thesis. Nijmegen, Netherlands, 1987.

Lange A de, Kauer JMG, Huiskes R: Kinematic behavior of the human wrist joint: a roentgen-stereophotogrammetric analysis. J Orthop Res 1985;3:56-64.

Linscheid RL: Kinematic Considerations of the Wrist. Clin Orthop 1986;202:27-39.

Logan SE, Vannier MW, Bresina SJ, Weeks PM: Kinematic analysis of human wrist motion using a six-degree-of-freedom digitizer and computer-assisted design system. Surg Forum Amer Coll Surgeons 1985;36:568-571.

Ruby LK, Cooney WP, An KN, Linscheid RL, Chao EYS: Relative motion of selected carpal bones: A kinematic analysis of the normal wrist. J Hand Surg 1988;13A:1-10.

Ruby LK, An KN, Linscheid RL, Cooney WP, Chao EYS: The effect of scapholunate ligament section on scapholunate motion. J Hand Surg 1987;12A(2 Pt 1):767-771.

Smith DK, Cooney WP, An KN, Linscheid RL, Chao, EYS: The effects of simulated unstable scaphoid fractures on carpal motion. J Hand Surg 1989a.;14A:283-291.

Smith DK, An KN, Cooney WP, Linscheid RL, Chao EYS: Effects of a scaphoid waist osteotomy on carpal kinematics. J Orthop Res 1989b.;7:590-598.

Talesnik J: Carpal kinematics. In Taleisnik J, The Wrist. New York, Churchill Livingstone, 1985,pp39-49.

Von Bonin G: A note on the kinematics of the wrist joint. J Anat 1919;63:259-262.

Weber ER: Concepts governing the rotational shift of the intercalated segment of the carpus. Orthop Clin N Amer 1984;15:193-207.

Wright RD: A detailed study of movement of the wrist joint. J Anat 1935;70:137-142.

Youm Y, Yoon YS: Analytical development in investigation of wrist kinematics. J Biomech 1979;12:613-621.

Youm Y, McMurtry RY, Flatt AE, Gillespie TE: Kinematics of the wrist. I - An experimental study of radial/ulnar deviation and flexion/extension. J Bone Joint Surg 1978;60A:423-431.

Chapter 5

Force Analysis

F.W. Werner, K-N. An., A.K. Palmer, and
E.Y-S. Chao

Introduction

The amount of information known about the forces in the wrist joint is limited, but has been rapidly increasing over the past few years. These forces are a result of those externally applied and those exerted by the wrist and finger flexor and extensor muscle-tendon units. The combination of eight semi-rigid carpal bones located between the radius, ulna, and five metacarpal bones; compressive load bearing tissues varying in compliance from cartilage to the triangular fibrocartilage complex; ligaments spread throughout the wrist; and muscle forces being applied across the wrist joint, all add up to make the wrist one of the most complex joints in the body. The numerous wrist positions and motions only complicate the problem further. In addition, there is an interaction between the forces and motions at the wrist and those at the elbow and hand. The complexity of the wrist has required researchers to develop new methods of theoretically and experimentally estimating the forces in the joint. At this point, the results obtained still only reflect what can either be analytically modeled or experimentally measured from in vitro cadaver experiments.

The methods used to determine the forces in the wrist include using free body diagrams, a rigid body spring model, force transducers, pressure sensitive film, pressure transducers, and strain gauges. Some of these methods, such as the use of pressure sensitive film and strain gauges will be discussed in subsequent chapters. This chapter will first cover the methods and results of theoretical approaches. Next, the variety of experimental tools and methods will be described. The forces measured in the normal wrist will then be presented, followed by a discussion of the way these forces are altered by various clinical problems and surgical treatments.

Theoretical Methods And Results

A free body diagram can frequently be used to depict and analyze the forces acting on a structure. To illustrate the difficulty in using this method, let us consider a lateral view of the wrist, as shown in Figure 1. To simplify, only a two-dimensional force analysis is examined. If we assume the forearm is vertical, and the wrist is in extension and is being used to lift or hold a container, forces must be exerted by the wrist flexors. Assuming the center of rotation for flexion and extension is located at the proximal head of the capitate (Youm, et al., 1978), and that there are no dynamic effects, then we can sum the moments about the center of rotation and find:

$$F_{flexors} = (F_{container})X(R_{container})/(R_{flexors})$$

Even in this extremely simple model, there are a few problems. First, the center of rotation is not a fixed point; thus, as the hand moves, these moment arms vary as a function of the wrist flexion angle. Secondly, and more importantly, there is not just one flexor at the wrist. There are two primary wrist flexors: the flexor carpi ulnaris and the flexor carpi radialis; and then there are all of the finger flexors which can also play a role in causing flexion of the wrist. Thirdly, in order to have a controlled motion of the forearm, the extensors will act as antagonists and slightly resist the motion or slow it down. These factors add additional unknown variables to this illustration and lead us to a situation known as a statically indeterminate problem where there are more unknowns then there are equations. Although optimization schemes have been used successfully to help in the solution of this type of problem in the knee and hip, the complexity of the wrist has thus far prevented their use here.

A three dimensional free body diagram requires even more assumptions. In particular, one would have to clarify the relative role of each muscle tendon. For example, the extensor carpi ulnaris functions as both a wrist extensor and ulnar deviator. During wrist extension, it still has a role in maintaining a chosen radial/ulnar deviation position. Much more research is needed to define the relative role of these and other structures in the wrist before this free body diagram technique can be used.

Another analytical approach is the rigid body spring model which represents the individual bones as rigid bodies and utilizes a number of compressive and tensile springs to connect the different structures. This method was used (Tsumura, et al., 1982; Tsumura and Himeno, 1983; Tsumura, et al., 1987) to examine several treatments for Kienböck's disease. The general development and validation of this method has been detailed recently by An et al. (1990).

The rigid body spring model (RBSM) can determine the two-dimensional joint contact pressure for any shape of articular surface and loading condition. As described by Horii et al. (1990), for use in the wrist, geometrical data is first obtained from contours of the carpal bones and intercarpal joint areas digitized from tomograms. A series of compression springs represent the reaction forces between the carpal bones. Tensile springs model 27 wrist ligaments. External loads are applied at the midaxis of the metacarpals and the carpal bones are allowed to displace and reach equilibrium. The joint forces can be tabulated for both the normal wrist and for variations in the normal anatomy due to trauma, disease, or surgery.

Using this model, Horii et al. (1990) examined the forces in the normal wrist, and, after simulating five treatments for Kienböck's disease (STT fusion, SC fusion, CH fusion, CH fusion combined with capitate shortening, and 4 mm of joint leveling). Tables 1A and 1B depict their findings based on a series of six arms for a total force of 143N applied to the metacarpals. In the intact wrist they found that 22% of the total force in the radial ulnar carpal joint is transmitted through the ulna (14% through the ulnolunate joint, and 8% through the ulnotriquetral) and 78% is transmitted through the radius (46% through the scaphoid fossa, 32% through the lunate fossa). At the midcarpal joint, the scaphotrapezial joint transmitted 31% of the total applied force, the scaphocapitate 19%, the lunocapitate 29%, and the triquetral hamate 21%. All of these results are for the wrist in a neutral position and are depicted graphically in Figure 2. The three limited intercarpal fusions were found to not reduce the compressive forces at the radiolunate (RL) joint by more than 15%. They found that the combined CH fusion and capitate shortening markedly reduced the RL forces at the expense of increasing the forces at the adjacent joints. A 4 mm ulnar lengthening (or 4 mm radial shortening) caused a 45% decease in the forces at the RL fossa, without adversely affecting the other joints.

Previous work by Tsumura et al. (1982; 1987) using the RBSM concept, showed the beneficial effect of the STT fusion and ulnar lengthening in decreasing the forces at the radiolunate joint. Tsumura and Himeno (1983) have also simulated carpal injury - perilunar instability by applying a horizontal load to the trapezoid with the wrist in ulnar deviation.

TABLE 1A. Intercarpal Joint Force Transmission (N) at the Radioulnocarpal Joint (Based on Rigid Body Spring Model)

	Ulno-Triquetral	*Ulno-Lunate*	*Radio-Lunate*	*Radio-Scaphoid*
Intact	12	23	52	74
STT Fusion	11	22	49	77
SC Fusion	11	22	46	80
CH Fusion	8	25	55	72
Capitate Shortening	26	23	18	93
4 mm Leveling	21	35	25	78

TABLE 1B. Intercarpal Joint Force Transmission (N) at the Midcarpal Joint. (Based on Rigid Body Spring Model)

	Triquetral-Hamate	*Luno-Capitate*	*Scapho-Capitate*	*Scapho-Trapesial*
Intact	36	51	32	51
STT Fusion	35	48	40	--
SC Fusion	34	45	--	53
CH Fusion	20	60	38	46
Capitate Shortening	90	--	16	87
4 mm Leveling	38	44	39	51

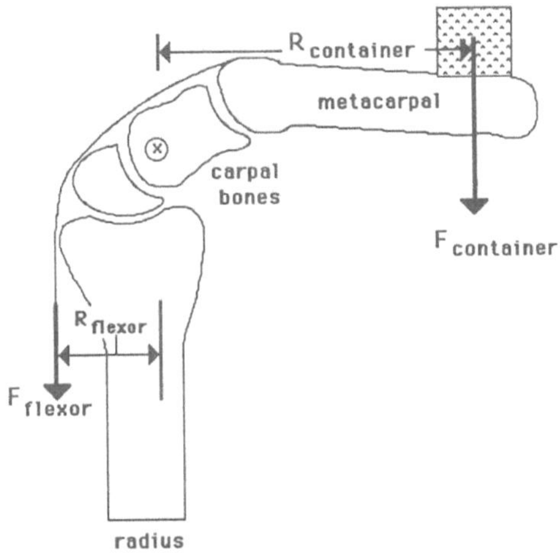

FIGURE 1. Two dimensional (sagittal plane) free body diagram of the forces transmitted through the wrist joint in extension. The center of rotation is shown as a circle with a "x" inside of it. The force of the container on the metacarpals is being counteracted by the flexor force.

The RBSM technique has been used by Garcia-Elias et al. (1989) to examine the stabilizing structures of the transverse carpal arch. In this model of the carpal arch a compression load was applied in the dorsopalmar direction in the intact carpus and after the sectioning of different ligaments. It was found that the intact carpus has a stiffness of 232 N/mm, primarily supported by the flexor retinaculum, and the palmar hamate-capitate, the palmar capitate-trapezium, and the dorsal hamate-capitate ligaments. The compressive and tensile intercarpal forces for the intact carpus are shown in Figure 3. Major changes in the ligament forces and carpal stiffness occurred with isolated removal of the palmar hamate-capitate ligament or with removal of one palmar intercarpal ligament combined with removal of the flexor retinaculum.

New analytical models of the wrist will require the use of additional quantitative data on the properties of different soft tissues in the wrist such as different wrist ligaments (which will be discussed in Chapter 8), and geometrical information such as the moment arms of each of the muscle tendons passing over the wrist. Studies by Ohnishi et al. (1991), who has already examined the moment arms of twenty-four finger and wrist tendons, supplements the work by Brand et al. (1981) in this area and will be discussed in Chapter 9.

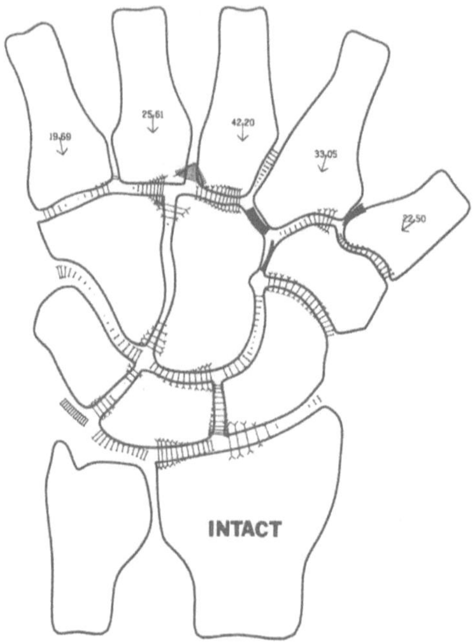

FIGURE 2. Relative compressive forces between adjacent bones in the wrist as calculated using the rigid body spring model. The wrist is intact and in a neutral position. Forces are applied as shown through the metacarpal bones (credit J Hand Surg 1990;15A:393-400; by Horii et al.).

Experimental Techniques

Numerous experimental techniques have been used in quantifying the forces in the wrist joint. The use of pressure sensitive film has been used at the radioulnar-carpal joint, the distal radial ulnar joint, and at the midcarpal joint. This material produces a pressure mapping, which, if integrated over the region of interest can be resolved into forces. The use of pressure sensitive film is primarily discussed in Chapter 6. Strain gauges have been bonded onto a number of locations in the body, including bones in the wrist joint. Their use is discussed in Chapter 7. Thin pressure sensitive conductive rubber transducers have been placed between articulating joints, particularly at the radioulnar-carpal joint. These transducers allow dynamic output of pressures at a limited number of locations on the articulating surface. Force transducers have been used to measure the forces being transmitted through a bone, such as the radius and ulna, or in one study in the scaphoid (Shaw,

1987). In vivo determination of forces in the tendons crossing over the wrist joint has been accomplished by transducers connected to the tendons. In vitro forces have been measured by transducers in series with the tendons.

Several methods have been used to apply loads to the forearm. In Figure 4, forces are applied to the major wrist flexor and extensor tendons, frequently by weights. Another method is shown in Figure 5, where the metacarpals are potted, and the force is applied directly from above. As discussed in Chapter 6, and here, similar results have been obtained by the two methods.

Most experimental studies have been performed with the forearm in a neutral supination/pronation position, and the wrist in neutral radial/ulnar deviation and neutral flexion/extension. A few studies have looked at limited combinations of these positions or a matrix of all positions.

Several investigators (Palmer and Werner, 1984; Werner, et al., 1986; Trumble, et al., 1986; Patterson, et al., 1991) have mounted load cells directly to the radius and ulna. Figure 4 illustrates the method used by Palmer and Werner (1984) where load cells are mounted onto the distal intact radius and ulna. After they are attached, an intermediate slice of bone is removed, and the portion of the interosseous membrane distal to the load cells is removed. If this portion of the interosseous membrane is not removed, force transfer can occur from the radius to the ulna, causing the ulnar load cell to measure higher forces. Maintenance of the original ulnar variance (relative height of the ulna with respect to the radius) is critical.

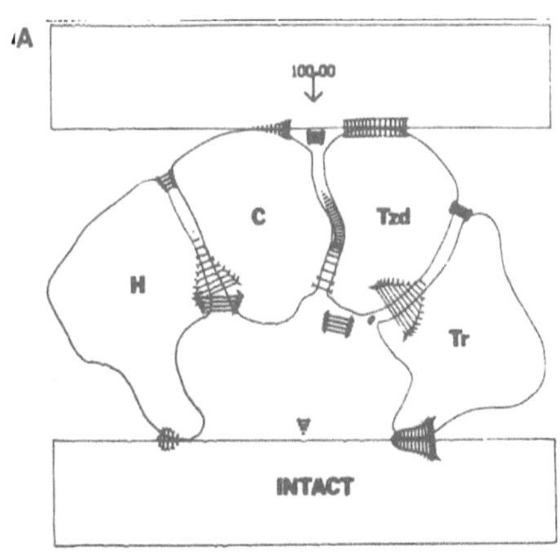

FIGURE 3. Compressive and tensile forces in the intact transverse carpal arch due to an applied 100N load (credit J Orthopaedic Res 1989;7:738-743; by Garcia-Elias et al.).

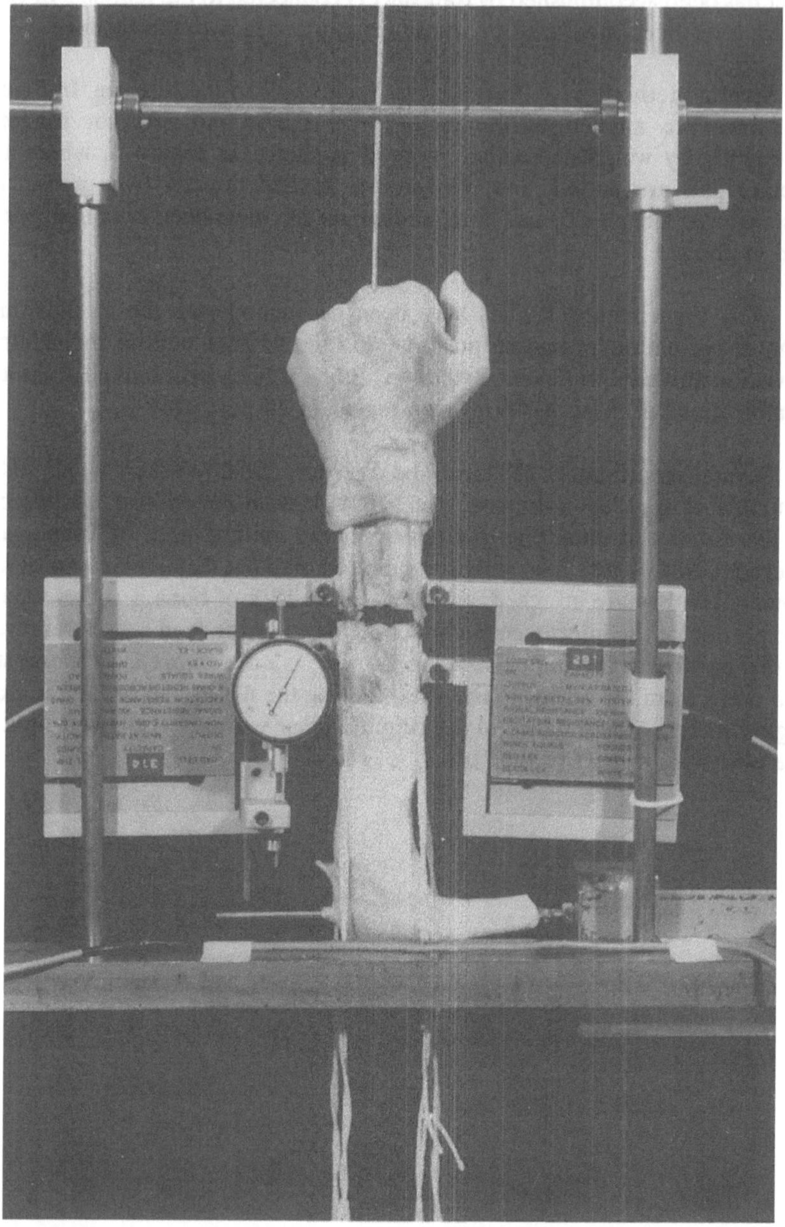

FIGURE 4. Experimental loading configuration of Palmer and Werner. Forces are applied to the wrist flexors and extensors, with the elbow in 90° of flexion. The forces transmitted through the distal radius and ulna are measured by load cells attached to those bones.

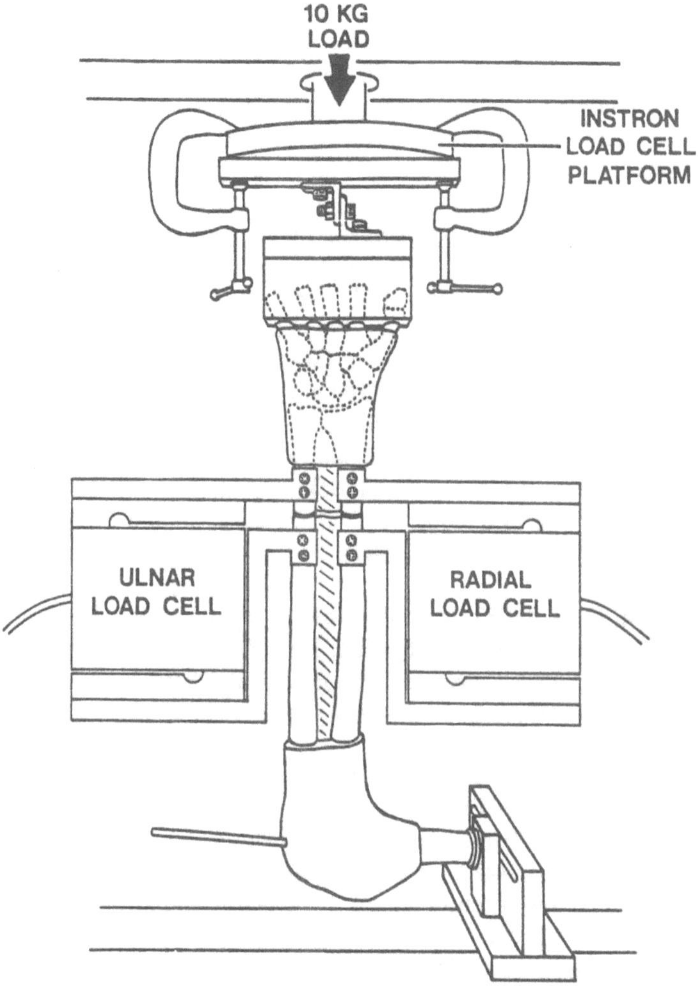

FIGURE 5. Experimental loading configuration of Trumble et al. Forces are applied through the potted metacarpals. Load cells are used to measure the forces transmitted through the radius and ulna (credit J Hand Surg 1987;12A:196-202, by Trumble et al.).

Normal Wrist Forces

In the intact forearm, in neutral forearm and wrist position, approximately 82% of the force is transmitted through the distal radius, and 18% through the distal ulna (Palmer and Werner, 1984). Similar findings were reported by

Trumble et al. (1987) (17% force through the ulna). Patterson et al. (1991) have recently reported that force distribution between the radius and ulna was 62% and 38% respectively in the neutral forearm position. A possible explanation for the difference in the force percentages, is that in the Patterson study, none of the interosseous membrane was removed. However, their load cells were mounted distal to the major band of the interosseous membrane.

The force distribution between the radius and ulna for a variety of wrist positions is shown in Table 2. With ulnar deviation (Werner, et al., 1986), or forearm pronation (Ekenstam, et al., 1984), there is an increase in the force transmitted through the distal ulna. There is a corresponding decrease in the ulnar force with radial deviation and forearm supination. One would expect with ulnar deviation an increase in the ulnar force. In pronation, the increase in ulnar force is thought to occur due to the increase in the ulnar variance with pronation. Based on the measurement of ulnar variance in fourteen cadaver forearms, Palmer et al. (1982) have shown with 60° of pronation there is an increase of 0.7 mm of ulnar variance. Similar findings have been demonstrated by Epner et al. (1982).

Trumble et al. (1987) demonstrated a slight gradual increase in ulnar force with increasing wrist extension. By 25° of extension, the ulnar force reached 23.6%, as compared to 17% at neutral. They found with 5° of wrist flexion, the force decreased to 14.9%.

Horii et al. (1990), at the Wrist Biomechanics Workshop at the Mayo Clinic, presented their findings using pressure sensitive conductive rubber transducers (PSR). These sensors are thin (0.9 mm) and flexible, can be custom made, and do not have to be replaced for each loading configuration. Figure 6 shows the positioning of 12 sensors and the relative pressure distribution in the radial ulnar carpal joint for the wrist in a neutral position under 100N of axial load. The greatest pressures were typically in the radial scaphoid fossa. The resultant forces in each fossa were calculated by summing up the products of pressure and the area of each sensor within that region. They found that 50% of the force was transmitted through the radial scaphoid fossa, 35% through the radial lunate fossa, and 15% through the triangular fibrocartilage complex (TFCC).

Horii et al. using the PSR sensors also examined the effect of different wrist positions on the forces in the wrist joint. In ulnar deviation, pressure in the lunate fossa increased, while that in the TFCC decreased. The changes in force transmission are shown in Figure 7. In ulnar deviation, the force through the lunate increased to 50%, and the force through the TFCC decreased to 8%. In extension, the only statistically significant change was an increase in the force in the lunate fossa to 52%.

TABLE 2. Percent Force Distribution in the Intact Wrist

	Radius	*Ulna*
Neutral Position	81.6	18.4
Ulnar Deviation	71.6	28.4
Radial Deviation	87.2	12.8
Forearm Pronation	63.0	37.0
Forearm Supination	86.0	14.0

Werner et al. (submitted) correlated the force distribution in the distal ulna to seven radiological measurements of the distal ulna (ulnar variance, radial tilt of the distal radius, palmar tilt of the distal radius, angulation of the lunate fossa, carpal height, carpal ulnar distance, and ulnar head inclination). With the possible exception of ulnar variance, none of these measurements is related to the amount of force transmitted through the distal ulna. The relationship between the force transmission and ulnar variance had the highest regression value (0.44) of any of the parameters, but this only suggests there may be a relationship, if any at all. These findings suggest that for the clinical setting, one cannot necessarily conclude that because a patient has a positive ulnar variance that this patient has a larger percentage of load transmitted through his or her ulna.

One of the frequently overlooked structures in the wrist joint is the triangular fibrocartilage complex (TFCC). Palmer et al. (1984) have demonstrated there is an inverse linear relationship between the thickness of the TFCC and the ulnar variance. In effect, the TFCC acts as a spacer, filling up the void between distal ulna and the carpus. This and other soft tissue factors may have an important role in how force is transmitted through the wrist joint.

Data presented at the Wrist Biomechanics Workshop at the Mayo Clinic by Short et al. (1990; 1991) demonstrated a good correlation between forces measured using load cells and forces obtained from pressure sensitive film. In a series of six arms in a neutral forearm and wrist position, the ulna transmitted 22.9% of the force. Integration of the pressure mapping across the radial ulnar carpal joint, indicated that 20.5% of the force occurred in the ulnar carpal fossa, 29.3% in the radial lunate, and 50.2% in the radial scaphoid.

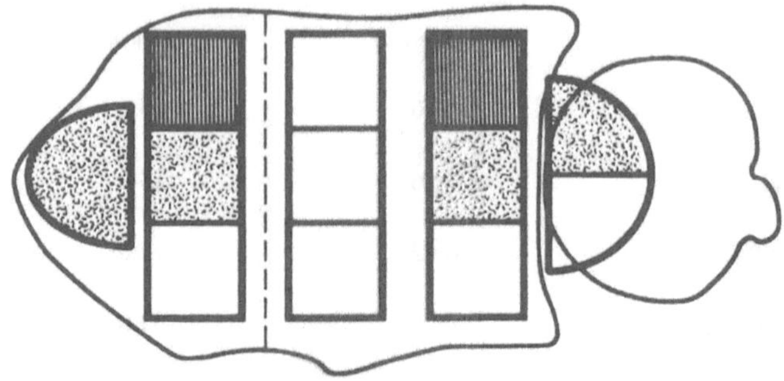

FIGURE 6. Positioning of 12 pressure sensitive conductive rubber sensors used in the radial ulnar carpal joint. The relative pressure distribution within a neutrally positioned wrist is shown, with darker regions having more pressure (credit Horii et al.).

Thus, the findings by An et al. (1990) using the RBSM, by Horii et al. (1990) using the pressure sensitive rubber transducer, and the findings by Short et al. (1990) based on pressure sensitive film are in agreement with the load cell based findings that approximately 20% of the force is transmitted by the distal ulna and 80% through the radius in the neutral position.

Direct in vivo measurement of the forces in the tendons passing over the wrist joint have been made using transducers, such as the one described by An et al. (1990). Rodgers et al. (1989) performed an intraoperative assessment of muscle performance during total wrist arthroplasty to determine whether tendon lengths should be altered at the time of surgery. Mendelson et al. (1988) assessed the force contribution of the extensor carpi radialis brevis (ECRB) and extensor carpi radialis longus (ECRL). They wanted to verify that if the ECRL was used as part of a tendon transfer, that the ECRB would provide sufficient force to allow extension of the wrist. The average peak force, whether due to voluntary contraction or due to electrical stimulation, was 73N. In most patients, the ECRL developed more force than the ECRB.

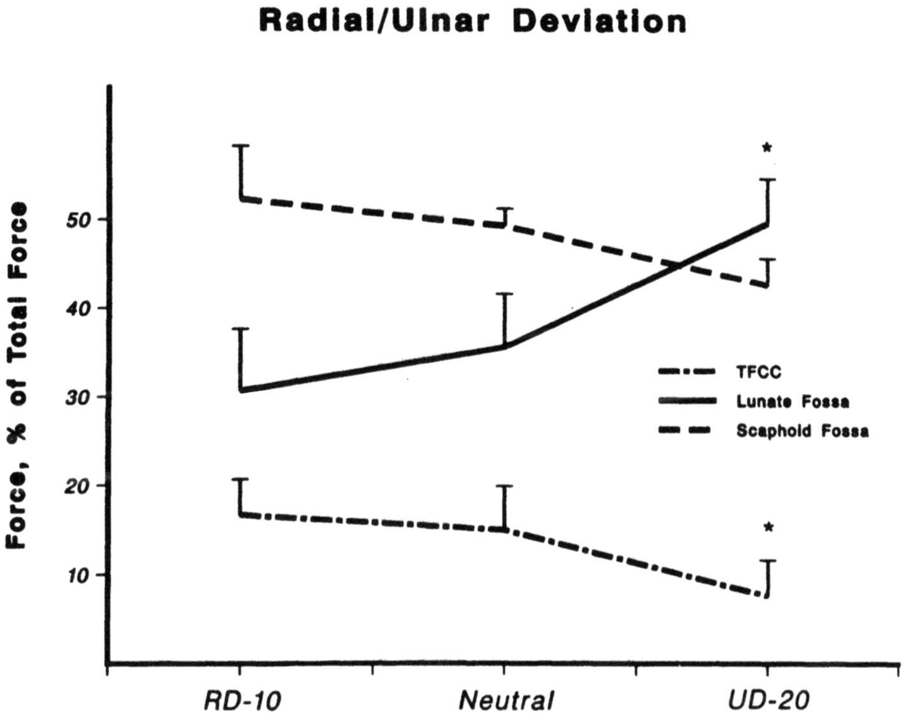

FIGURE 7. The relative force transmission through the radial ulnar carpal joint, using the pressure sensitive conductive rubber sensors in a neutral wrist position and in radial and ulnar deviation (* = statistically differently from neutral position, $p<0.05$) (credit Horii et al.).

Dynamic measurement of wrist tendon forces has been accomplished using a mechanical wrist joint motion simulator. Tong et al. (1986) reported on the development of a servohydraulic simulator that would cause planar flexion/extension and planar radial/ulnar deviation in cadaver wrists by applying forces directly to the wrist flexor and extensor tendons. The tendon forces and wrist flexion/extension and radial/ulnar deviation positions, recorded by a computer, are used as input for the position feedback control algorithm which determines the amount of contraction needed in each tendon to move the hand incrementally from the actual position to the desired position. Position feedback control is used to solve the indeterminate problem of having more than one combination of tendon forces that could achieve a desired position or motion. Antagonist tensions are maintained in the wrist tendons. Table 3 shows the average maximum tensions that occurred during cyclic flexion/extension and radial/ulnar deviation tests of five intact cadaver forearms. These results are based on a constant antagonistic force of 8.9N and for a total of fifteen cycles. During the cyclic 28° flexion to 28° extension motion, the largest force was consistently the ECR (ECRB and ECRL held together). To achieve a cyclic 14° radial deviation to 14° ulnar deviation motion, greater forces were required from the wrist extensors than the flexors.

Additional data from this wrist joint simulator was presented by Werner et al. (1989). After using the simulator to move the wrist to a specified position, such as 28° of flexion, the forces were measured to maintain that static position. Table 4 shows the average relative forces required to maintain several positions, for eight fresh cadaver arms, normalized to the force in the ECR (again the combination of the ECRB and ECRL). In nearly all cases, the ECR is the largest. As expected, the relative contribution of the extensor carpi ulnaris (ECU) increased with ulnar deviation, decreased with radial deviation, increased with extension, and decreased with flexion. The dual role of each tendon is illustrated in radial deviation for the flexor carpi ulnaris (FCU) and flexor carpi radialis (FCR). Here, there is an increase in both the FCU and FCR. The unexpected increase in FCU force is perhaps required to maintain the neutral flexion/extension position.

TABLE 3. Average Maximum Dynamic Muscle Tensions (N) During Cyclic Motions. (Based on a Wrist Joint Motion Simulator)

	ECU	*ECR*	*FCU*	*FCR*
Flexion/Extension Motion	39.1	50.7	44.0	32.9
Radial/Ulnar Deviation Motion	34.2	33.4	23.6	22.2

TABLE 3. Legend:

ECU = Extensor Carpi Ulnaris
ECR = Combination of Extensor Carpi Radialis Brevis and Extensor Carpi Radialis Longus
FCU = Flexor Carpi Ulnaris
FCR = Flexor Carpi Radialis

TABLE 4. Relative Tendon Forces Required to Hold Indicated Static Positions (% of ECR Force). (Based on a Wrist Joint Motion Simulator)

Wrist Position	*ECU*	*ECR*	*FCU*	*FCR*
Neutral Flex/Ext	51.4	100	43.7	25.6
28° Flexion	35.6	100	34.5	21.5
28° Extension	91.9	100	106.6	65.8
14° Radial Dev.	35.0	100	54.8	32.7
14° Ulnar Dev.	61.8	100	43.5	23.5

TABLE 4. Legend:

ECU	=	Extensor Carpi Ulnaris
ECR	=	Combination of Extensor Carpi Radialis Brevis and Extensor Carpi Radialis Longus
FCU	=	Flexor Carpi Ulnaris
FCR	=	Flexor Carpi Radialis

Experimental Forces In The Diseased, Traumatized, Or Surgically Treated Wrist

The experimental determination of the forces in the wrist joint has primarily been based on elderly fresh frozen cadaver specimens due to their availability. Most of these specimens are relatively normal and do not reflect the diseased, traumatized, or clinically treated wrists. These abnormal conditions are usually simulated in the experimental setup and thus do not include any bone or soft tissue changes that can occur with time in the patient.

Clinical symptoms in the region of the distal ulna may warrant surgical intervention to the triangular fibrocartilage complex (TFCC) or the distal ulna. Palmer and Werner (1984) demonstrated that complete excision of the TFCC reduces the force transmitted through the distal ulna from 18.4% to 6.2%. This finding and others regarding its stabilizer role (Palmer and Werner, 1981) support the importance of keeping the TFCC whenever possible. This research group also showed that resection of the distal ulna (as expected) eliminated any force transmission through it. Traumatic tears to the TFCC may cause impingement or pain, and thus require a partial resection of its articular disk component. A study to examine the consequences of its incremental removal (Figure 8) was performed by Palmer et al. (1988). Table 5 shows how the force transmitted through the distal ulna decreases with removal of thirds of the TFCC. Only when 2/3 or more of the articular disc portion of the TFCC is removed was the force statistically decreased. These results support, from a biomechanical standpoint, partial, instead of total, excision of the central portion of the TFCC, for treatment of a central TFCC perforation.

The ulnar impaction syndrome is also a cause of wrist pain and is characterized by the deterioration of the TFCC. It is usually associated with wrists with a positive ulnar variance. Wnorowski et al. (1990) investigated the use of an arthroscopic wafer-type distal ulna resection procedure to determine how much of the distal ulna should be removed to relieve the pain caused by ulnar load. In this study, only cadaver forearms having a positive ulnar variance were used. Twelve conditions were examined: the normal wrist, one with resection of the horizontal portion of the TFCC, with the removal of incremental thirds of cartilage going ulnarly, with removal of incremental thirds of a 1 mm depth of subchondral bone, with incremental removal of a second 1 mm depth of subchondral bone, and with the removal of a third 1 mm depth of bone. The percentage force transmitted through the distal ulna decreased from 20.8% for the normal condition to 16% with excision of the TFCC. Removal of the peripheral 1/3 width of articular cartilage decreased the force to 13.6%. After the 3 mm of bone was resected, the force decreased to 10.8%. They concluded that this method has a biomechanical benefit in unloading the ulnocarpal articulation. The amount that should be removed appears to depend more upon the stage of TFCC degenerative pathology (Wnorowski, et al., 1990) than the amount of positivity of ulnar variance.

TABLE 5. Percent Force Transmitted Through the Distal Ulna

Amount Removed of the Articular Disk of the Triangular Fibrocartilage Complex:

Intact	*1/3 Removed*	*2/3 Removed*	*All Removed*
17.6	16.1	13.4	8.0

In the treatment of Kienböck's disease, a number of surgical procedures have been used. Most of these are intended to alter the relative length difference between the radius and ulna and thus unload the lunate. The analysis of some of these is also covered in Chapters 6 and 7. Even though several of the treatments reviewed (Werner, et al., 1988; Werner, et al., 1987) in Chapter 6, include data on the forces transmitted through the wrist, they are included in chapter 6 since they also have radial ulnar carpal joint pressure distribution findings based on pressure sensitive film. Of interest in these studies is the remarkable alteration in load distribution between the radius and ulna due to distal radial wedge osteotomies (Werner, et al., 1988) and limited intercarpal fusions (Werner, et al., 1987).

Two common joint leveling procedures, used for Kienböck's, are lengthening of the ulna or shortening of the radius. Figure 9 from the work of Palmer and Werner (1984) shows the relationship between ulnar force and experimentally caused changes in ulnar variance. A 2.5 mm ulnar lengthening will cause the radius to bear 58% of the total load, and the ulna 42%. Although not used for Kienböck's, a 2.5 mm ulnar shortening decreases the ulnar force to only 4%. When a TFCC excision is combined with a 2.5 mm ulnar lengthening, the ulnar force is 22%, but when combined with a 2.5 mm ulnar shortening, the ulnar force is just 3%. The effect of a 2.5 mm radial shortening was the same as a 2.5 mm ulnar lengthening.

The results of four total wrist implants on the forces transmitted through the major wrist extensors and flexors were studied by Werner et al. (1989). Each implant was implanted and tested in three fresh cadaver forearms. The implants tested were the Volz, Beckenbaugh, Hamas, and Taleisnik. A computer controlled servohydraulic wrist joint motion simulator was used to dynamically move each wrist in first cyclic planar flexion/extension, and then in cyclic planar radial/ulnar deviation. Data was recorded before and after an implant was in place for each of the twelve forearms.

A comparison of the forces required to maintain specified static positions, such as those listed in Table 4, showed that in order to maintain neutral flexion/extension after an implant was inserted, a smaller ECU force, a larger ECR force, and a smaller FCU force was required. To maintain radial deviation, larger radial deviator forces were required.

The effect of the four implants on the dynamic forces of each tendon while acting as an agonist during planar wrist motions is shown in Tables 6 and 7. In both motions, the implants, in general, required greater radial deviator forces and smaller or equivalent ulnar deviator forces than when the arms were intact. This in vitro study shows that each of the four implants tested alters the normal biomechanics of the wrist.

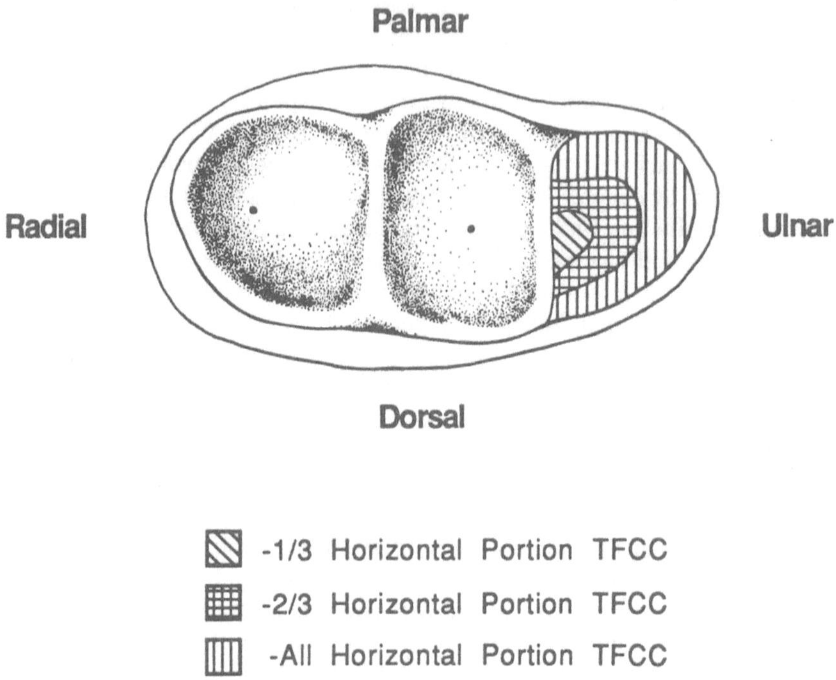

FIGURE 8. An artist's rendering of the experimental removal of 1/3, 2/3, and all of the horizontal portion of the triangular fibrocartilage complex (credit Palmer, The distal radial ulnar joint, Hand Clinics 1987;3:1).

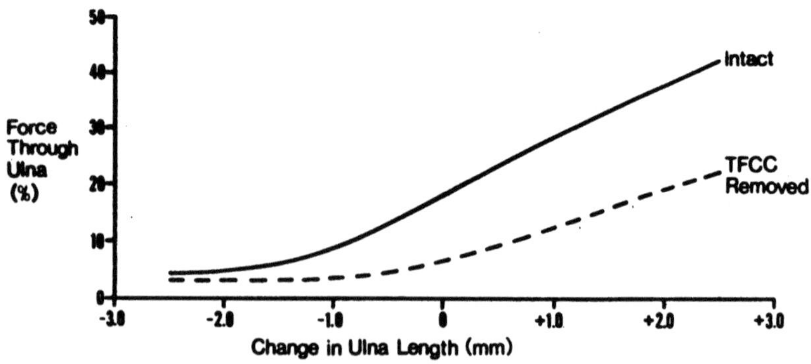

FIGURE 9. Relative force transmission through the distal ulna using load cells mounted to the radius and ulna. Shortening of the ulna by 2.5 mm results in a drop in ulnar load to 4%. Lengthening of the ulna by 2.5 mm results in an increase in ulnar load to 42%. Similar, though less dramatic, changes as seen in wrists tested after the triangular fibrocartilage complex has been removed (credit Palmer and Werner, Biomechanics of the Distal Radial Ulnar Joint, Clinical Orthopaedics 1984;187:26-35).

TABLE 6. Effect of Wrist Implant on Dynamic Tendon Forces for Planar Flexion/Extension Motion (% of Intact). (Based on a Wrist Joint Motion Simulator)

	ECU	*ECR*	*FCU*	*FCR*
Beckenbaugh	100	200	100	100 to 200
Hamas	100 to 200	100 to 200	100 to 300	100 to 300
Taleisnik	35	100	100	100
Volz	50 to 100	200 to 400	50 to 100	200

TABLE 6. Legend:

ECU = Extensor Carpi Ulnaris
ECR = Combination of Extensor Carpi Radialis Brevis and Extensor Carpi Radialis Longus
FCU = Flexor Carpi Ulnaris
FCR = Flexor Carpi Radialis

TABLE 7. Effect of Implant on Dynamic Tendon Forces for Planar Radial/Ulnar Deviation Motion (% of Intact). (Based on a Wrist Joint Motion Simulator)

	ECU	*ECR*	*FCU*	*FCR*
Beckenbaugh	35 to 50	200 to 300	50 to 100	200
Hamas	50	100	100	200
Volz	50 to 150	100 to 400	100	100 to 300

TABLE 7. Legend:

ECU = Extensor Carpi Ulnaris
ECR = Combination of Extensor Carpi Radialis Brevis and Extensor Carpi Radialis Longus
FCU = Flexor Carpi Ulnaris
FCR = Flexor Carpi Radialis

Much of what we know about the forces in the wrist joint has been experimentally determined and is concentrated at the radial ulnar carpal joint. Only recently, have researchers been quantifying the intercarpal joint and ligamentous forces. New analytical tools have opened up research areas previously unexamined, yet much is still unknown about the forces in the wrist joint.

References

An KN, Berglund L, Cooney WP, Chao EYS, Kovacevic N: Direct In Vivo Tendon Force Measurement System. J Biomech 1990;23:1269-1271.

An KN, Himeno S, Tsumura H, Kawai T, Chao EYS: Pressure Distribution on Articular Surfaces: Application to Joint Stability Evaluation. J Biomech 1990;23:1013-1020.

Brand PW, Beach RB, Thompson DE: Relative Tension and Potential Excursion of Muscles in the Forearm and Hand. J Hand Surg 1981;6:209-219.

Ekenstam FW, Palmer AK, Glisson RR: The Load on the Radius and Ulna in Different Positions of the Wrist and Forearm. Acta Orthop Scand 1984;55:363-365.

Epner RA, Bowers WH, Guilford WB: Ulnar Variance - The Effect of Wrist Positioning and Roentgen Filming Technique. J Hand Surg 1982;7:298-305.

Garcia-Elias M, An KN, Cooney WP, Linscheid RL, Chao EYS: Transverse Stability of the Carpus: An Analytical Study. J Orthop Res 1989;7:738-743.

Horii E, An KN, Chao EYS: Force Distribution Across the Wrist Joint using a Pressure Transducer. Wrist Biomechanics Workshop, Mayo Clinic, October 7, 1990.

Horii E, Garcia-Elias M, An KN, Bishop AT, Cooney WP, Linscheid RL, Chao EYS: Effect on Force Transmission Across the Carpus in Procedures Used to Treat Kienböck's Disease. J Hand Surg 1990;15:393-400.

Mendelson LS, Peckham PH, Freehafer AA, Keith MW: Intraoperative Assessment of Wrist Extensor Muscle Force. J Hand Surg 1988;13:832-836.

Ohnishi N, Ryu J, Colbaugh R, Rowen B: Moment Arms of the Prime Wrist Motors and Extrinsic Finger Motors at the Wrist. Transactions of the Orthopaedic Research Society, Anaheim, California, 1991;pp567.

Palmer AK: Triangular Fibrocartilage Complex Lesions: Classification and Treatment. J Hand Surg 1989;14A:594-606.

Palmer AK, Werner FW, Glisson RR, Murphy D: Partial Excision of the Triangular Fibrocartilage Complex. J Hand Surg 1988;13:403-406.

Palmer AK: The distal radial ulnar joint. Hand Clinics 1987;3:31-40.

Palmer AK, Glisson RR, Werner FW: Relationship Between Ulnar Variance and Triangular Fibrocartilage Complex Thickness. J Hand Surg 1984;9A:681-683.

Palmer AK, Werner FW: Biomechanics of the Distal Radioulnar Joint. Clin Orthop 1984;187:26-35.

Palmer AK, Glisson RR, Werner FW: Ulnar Variance Determination. J Hand Surg 1982;7:376-379.

Palmer AK, Werner FW: The Triangular Fibrocartilage Complex of the Wrist - Anatomy and Function. J Hand Surg 1981;6:153-162.

Patterson RM, Todd P, Viegas SF, McCarty P: Forearm Load Distribution. Ninth Annual Conference on Biomedical Engineering Research, Houston, Texas, February 7-8, 1991.

Rodgers MM, Barre PS, Zachary SV, Glaser RM: Intraoperative Assessment of Muscle Performance During Total Wrist Arthroplasty. Transactions of the American Society of Biomechanics, Burlington, Vermont, 1989;pp36-37.

Shaw JA: A Biomechanical Comparison of Scaphoid Screws. J Hand Surg 1987;12:347-353.

Short WH, Werner FW, Fortino MD, Palmer AK: Distribution of Pressures and Forces on the Wrist After Simulated Intercarpal Fusion and Kiènböck's Disease. Transactions of the Orthopaedic Research Society, Anaheim, California, 1991;pp568.

Short WH, Werner FW, Fortino MD, Palmer AK: Distribution of Pressures and Forces on the Wrist After Simulated Intercarpal Fusions and Kienböck's Disease. Wrist Biomechanics Workshop, Mayo Clinic, October 7, 1990.

Tong J, Werner FW, Somerset JH: Wrist Joint Motion Simulator. Transactions of the Orthopaedic Research Society, New Orleans, Louisiana, 1986;pp441.

Trumble TE, Glisson RR, Seaber AV, Urbaniak JR: Forearm Force Transmission After Surgical Treatment of Distal Radioulnar Joint Disorders. J Hand Surg 1987;12:196-202.

Tsumura H, Himeno S, An KN, Cooney WP, Chao EYS: Biomechanical Analysis of Kienböck's Disease. Transactions of the Orthopedic Research Society, San Francisco, California, 1987;pp208.

Tsumura H, Himeno S: Load Transmission and Injury Mechanism of the Wrist Joint. Biomechanics Symposium 1983;56:5-8.

Tsumura H, Himeno S, Kojima T, Kido M: Biomechanical analysis of Kienböck's Disease. Seikeigeka 1982;3:1399-1402.

Werner FW, Palmer AK, Fortino MD, Short WH: Force Transmission Through the Distal Ulna: Effect of Ulnar Variance, Lunate Fossa Angulation, and Radial and Palmar Tilt of the Distal Radius. Submitted to J Hand Surg.

Werner FW, Palmer AK, Tong J, Somerset JH: In Vitro Evaluation of Total Wrist Replacements. Transactions of the Orthopaedic Research Society, Las Vegas, Nevada, 1989;pp366.

Werner FW, Palmer AK, Utter RG: Distal Radial Wedge Osteotomy for the Treatment of Kienböck's Disease: A Biomechanical Study. Transactions of the Orthopaedic Research Society, Atlanta, Georgia, 1988;pp411.

Werner FW, Murphy D, Palmer AK: Silastic Synovitis Following Lunate Replacement: A Biomechanical Evaluation of Preventive Measures. Transactions of the Orthopaedic Research Society, San Francisco, California, 1987;pp414.

Werner FW, Glisson RR, Murphy D, Palmer AK: Force Transmission Through the Distal Radioulnar Carpal Joint: Effect of Ulnar Lengthening and Shortening. Handchirurgie 1986;18:304-308.

Wnorowski DC, Palmer AK, Werner FW, Fortino MD: The Arthroscopic Wafer Procedure: An Anatomic and Biomechanical Study. Transactions of the Orthopaedic Research Society, New Orleans, Louisiana, 1990;pp501.

Youm Y, McMurtry RY, Flatt AE, Gillespie TE: Kinematics of the Wrist. J Bone Joint Surg 1978;60:423-431.

Chapter 6

Joint Contact Area and Pressure

S.F. Viegas, R.M. Patterson, and F.W. Werner

Introduction

During the later half of the past decade a considerable amount of information on the biomechanics of the human wrist has been uncovered by measuring the contact areas and pressures in cadaver wrists in a variety of normal, simulated post-traumatic, and surgically treated conditions. The results of these studies have added to our knowledge base and provided clinically relevant information on the normal functional mechanics of the wrist as well as insight on and guidelines for the treatment of a number of different fractures and ligament injuries in and about the wrist.

A number of these studies used pressure sensitive film to measure forces generated in the wrist by some method of loading the extremity either axially through the skeleton or by means of loading the tendons. The resulting pressure sensitive prints were studied by some sort of analysis system.

There have been a number of attempts to measure the forces and pressures generated in the wrist. Different types of transducers have been placed within the wrist joint or on the carpal bones to assess force. Piezo film is one such type of transducer which uses piezoelectric material that deforms when force is applied to the film. It is very thin and has the capability to be reused and connected to a computer, however, it will only measure dynamic forces. Thus, Piezo film can not be used for static load applications.

Strain gauges are another type of force transducer that has been used. Their application to bone, however, is difficult and generally requires removal of all soft tissue and cartilage as well as drying and degreasing the area to which the strain gauge is to be attached. They measure strain only in the area upon which the strain gauge is directly attached, and since bone is not a homogenous material it may not be representative of the overall bone forces. Also, to accurately measure the principal strains and their orientation, at even one location on a bone, a combination of three gauges (called a rosette) must be used.

Stress paint and photoelastic coatings are other modes of detecting contact. These methods can detect shear stresses and are not typically used to detect contact areas or pressures. These types of coatings can detect contact areas but can not quantatively measure the contact pressure (Ishizuka and Takao).

One pressure sensitive film or measuring sheet which has been developed by Fuji Photo Film Co., Ltd is called, "Prescale". It has been available since April 1977 and was developed in response to an industrial need for measuring contact pressures. It is used in hydraulic presses, in the manufacturing of flanges, and most extensively in the gasket industry to detect leakage of oil. More recently, pressure sensitive film has been used in medical applications.

This pressure sensitive film consists of two pieces of film (A and C film) that when placed together can detect pressures. The A film has one side coated with randomly distributed microspheres that burst under various predetermined pressures. The C film is a developer that turns a red color when it comes into contact with the material released by the A film. Fuji A film comes in four pressure grades, super low (71-284 PSI [0.5-2.0 N/mm2]), low (284-994 PSI [2.0-6.9 N/mm2]), medium (994-3551 PSI [6.9-24.5 N/mm2]), and high (3551-9943 PSI [24.5-68.5 N/mm2]).

Fuji film is thin (0.28 mm) and has an immediate observable pressure distribution pattern that is quantifiable. It has certain characteristics and limitations, however, and care must be taken when utilizing the film.

The film must remain dry since any contact with liquids or chemicals will distort the color intensity and therefore the measurement of force. Thus, for use in many biomechanical studies, the film must be sealed. An adhesive tape or sealant can be used to seal the transducer. An alternative to sealing the film is to thoroughly dry the joint surface on which the film will be used. Another characteristic of the film is that the print color fades with time and exposure to light. Fading is reduced if clear tape is placed over the print. The rate of fading, however, is not documented and analysis of the print should be performed immediately. Also, different boxes of film often have a slightly different color of developer or "C" film; some appearing red and others appearing more orange or purple.

The Fuji company states that the film requires no preparation for use, other than simply to cut the film into the required shape. In practical terms, this process is not always simple, particularly when the shapes required are complex, as in some human joints. The film can only be used once, and a new transducer must be made for each test. One time saving solution has been to fabricate a die to be used with a punch press to cut the A and C film in the shape of the custom die. The A and C film must still be superimposed, sealed or taped, and trimmed by hand. If it is desired to not freeze the arm while waiting for a die to be made, individual films can be cut out by hand. In either case, it is important that the transducer only covers the articulating surfaces of the joint.

This film measures the peak pressures between two surfaces produced at anytime during a loading regime. A pressure versus time plot can not be generated from pressure sensitive film prints. Care should also be taken during transducer preparation. Artifacts can be induced on the film from the factory or from making transducers, especially when working with the super low film. Shear forces are detected on the film and cause increased color density and false force readings (Rudert, et al., 1988). Wrinkling is another artifact that occurs when trying to fit the flat piece of film into a space. The film will wrinkle or fold on itself, causing false color lines radiating from the contact area (Rieck, et al., 1984). Analysis should always address this type of error. Investigators have found that as long as their analysis routine is consistent, whether including or excluding wrinkling artifacts, the data is not significantly influenced by the presence of this artifact.

A densitometer can measure the pressure generated by pressure sensitive film. This device analyzes a 3 mm diameter area providing a gauge measurement of the pressure. This process can be time consuming if the size and number of prints one wants to analyze are large.

A calibration of the film is needed to convert the color density to a known pressure. Since the color density on a film can be affected by changes in temperature and humidity, as well as which developer film is used, a calibration of the film should be done each day the film is used. Werner, et al. (1989) uses a method where seven known pressures are applied to a strip of film for subsequent analysis.

Fuji recommends loading the film for two minutes. Studies performed suggest that the film response becomes independent of time after thirty seconds (Rudert, et al., 1988). In the past Fuji has also recommended either a fast loading rate (less than 5 seconds) or the longer time of two minutes. It is still possible to use the shorter loading time if the calibration pressures are applied for the same duration of time and at the same rate of loading and unloading.

Viegas et al. (1987) and Palmer et al. (1984) have developed systems in which the low, medium, and high pressures can be calculated and stored. With the use of a video camera, light stand, videodigitizing board and computer, each print is stored as a graphics image. The Fuji prints are stored in the computer with calibration prints as soon as possible after the print has been generated. It can then be analyzed at any time. The calibration prints provide an accurate comparison between color density and actual pressure. An analysis program can then compare the grey levels of the print of the pressure calibrations videodigitized with each print, and the data can be stored for later statistical analysis.

Fuji prescale pressure sensitive film does have its limitations, however, it appears to be one of the easiest, most reliable, and most cost effective ways to measure intra-articular pressures.

Normal Wrist

Studies of the normal biomechanics of the proximal wrist joint have determined that the scaphoid and lunate bones had separate, distinct areas of contact on the distal radius/triangular fibrocartilage complex surface (Viegas, et al., 1987). Viegas et al. loaded the wrist with 23 lbs through the 2nd and 3rd metacarpals and found that the contact areas were localized and accounted for a relatively small fraction of the joint surface (average of 20.6%) regardless of wrist position (Figure 1). The contact areas shifted from a palmar location to a more dorsal location as the wrist moved from flexion to extension (Figure 2). Overall, the scaphoid contact area was 1.47 times that of the lunate and it was generally greatest with the wrist in ulnar deviation, i.e., with the scaphoid horizontally oriented. The scapholunate contact area ratio generally increased as wrist position changed from radial to ulnar deviation and/or from flexion to extension. Average high contact pressures within the scaphoid and lunate fossae varied with joint position; however, they were fairly low (average= 10 N/mm2, range 4 N/mm2 to 31.4 N/mm2 for an applied functional load of 103N (23 lbs). The intercentroid distance (scaphoid/lunate) averaged 14.9 mm and ranged from 10 mm to 20 mm depending on joint position (Viegas, et al., 1987).

Palmer et al. (1984) also studied pressures in the proximal wrist joint. They loaded the wrist with 20 lbs through the extensor and flexor tendons of the wrist and found that there are three distinct areas of contact: the ulnolunate, radiolunate, and radioscaphoid. They determined that the peak articular pressure in the ulnolunate fossa is 1.4 N/mm2, in the radiolunate fossa is 3.0 N/mm2, and in the radioscaphoid fossa is 3.3 N/mm2.

Viegas et al. (1989) found that when normal wrists were loaded over a range of weights, a non-linear relation was discovered between increasing loads and the overall contact areas (Figure 3). The general distribution between the scaphoid and the lunate contact areas was consistent at all loads tested, with 60% of the total contact area involving the scaphoid contact area and 40% involving the lunate contact area. Loads greater than 46 pounds were not found to significantly increase the overall contact areas, implying that the cartilage of the wrist joint was maximally compressed at loads of this magnitude (Figure 4). At loads higher than 46 pounds it appeared that average high pressures increased in a direct correlation with the increase in weight (Figure 5). The overall contact area, even at the highest loads tested, were not more than 40% of the available joint surface.

The wrist was also tested with a variety of load paths (i.e. two metacarpals, five metacarpals and tendons). This work implied that the distal carpal row acts functionally as a single unit in load transfer (Viegas, et al., 1989), which is consistent with its kinematic characteristics (Ruby, et al., 1988; Berger, et al., 1982; Lange, et al., 1985).

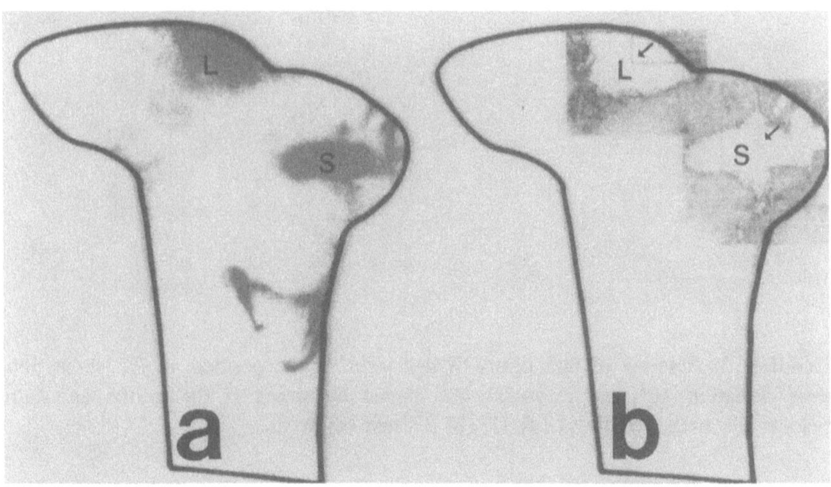

FIGURE 1: An example of the transducer print (a) and videodigitized computer analyzed printout (b) of a left wrist showing the scaphoid contact area (S) and lunate contact area (L), with their respective pressure centroids (arrows) located on the printout.

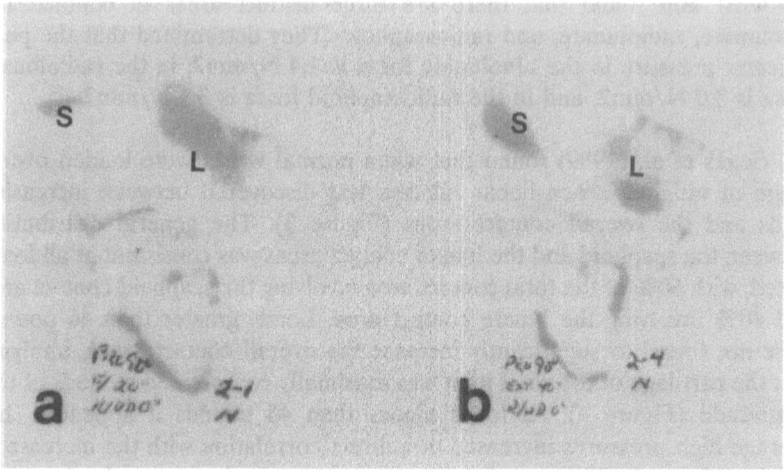

FIGURE 2: Examples of the typically palmar location of the contact areas of a right wrist with the wrist positioned in 20° of flexion (a) and the typically dorsal location of the contact areas with the wrist positioned in 40° of extension (b). (Scaphoid contact area [S], lunate contact area [L].

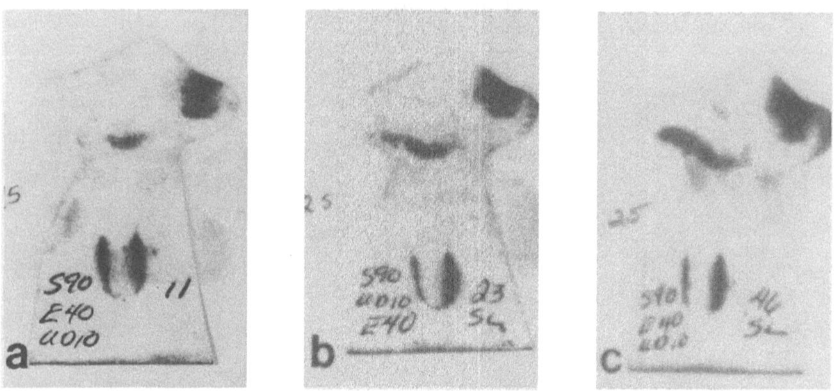

FIGURE 3: A series of film prints of one wrist in the position of full supination, ulnar deviation 10°, and extension 40° loaded by means of the second and third metacarpals with weights of (a) 11, (b) 23, and (c) 46 lb.

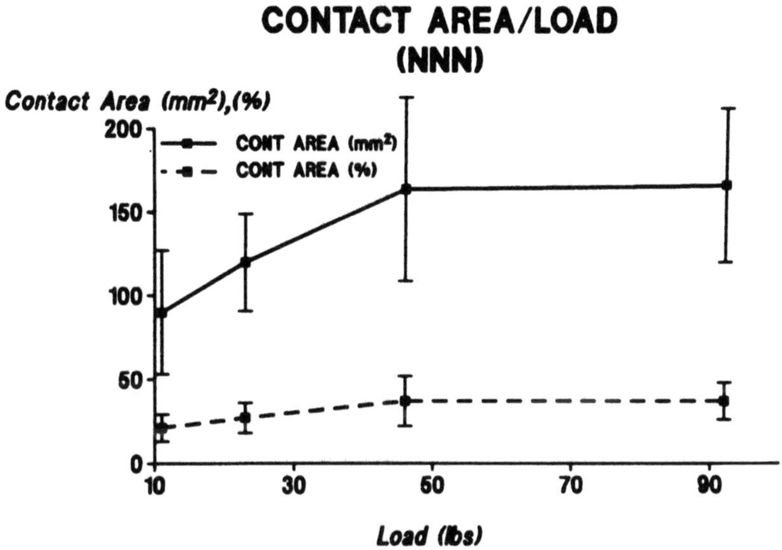

FIGURE 4: The non-linear relation between the contact area and the load, and the contact area normalized as a percentage of the available joint surface and the load of the position of neutral pronation/supination, neutral radioulnar deviation, and neutral flexion/extension of all wrists for loads of 11, 23, 46, and 92 lb.

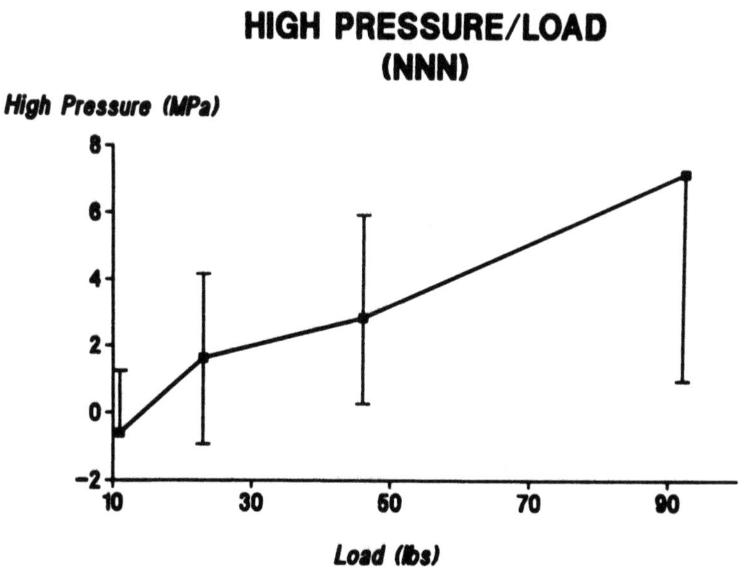

FIGURE 5: The average high pressure of all wrists in the position of neutral pronation/supination, neutral radioulnar deviation, and neutral flexion/extension in relation to loads of 11, 23, 46, and 92 lb.

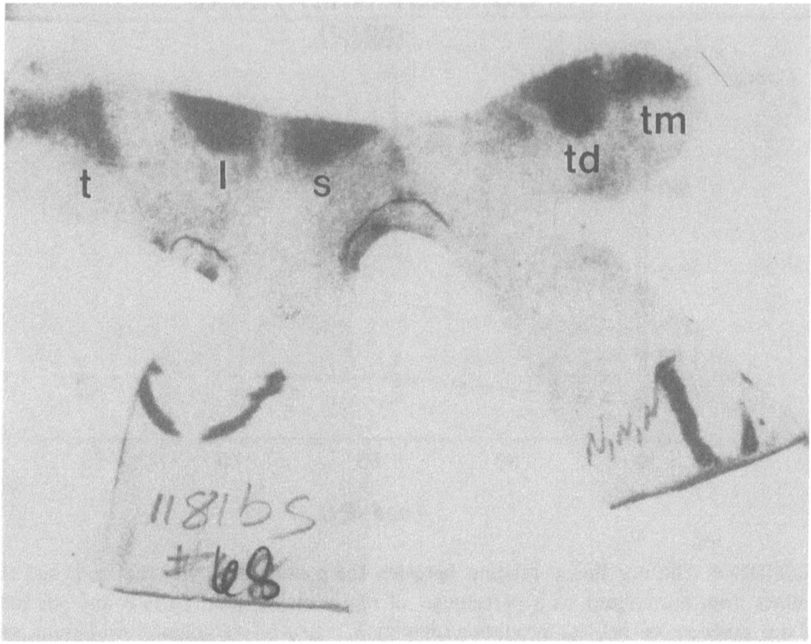

FIGURE 6: A print of the contact areas on the proximal aspect of the midcarpal joint of a left wrist loaded with 118 lbs, showing the trapezium and triquetrum contact on the scaphoid, the capitate contacts, one area on the scaphoid and one area on the lunate and the contact area of the hamate on the triquetrum.

There has been a limited amount of work performed in the midcarpal joint. This joint poses a problem because of its irregular shape. Viegas et al. (1990) have studied this joint and found four general areas of contact: the scapho-trapezium-trapezoid (STT), the scapho-capitate (SC), the capito-lunate (CL), and the triquetro-hamate (TH) (Figure 6). The high pressure contact area accounted for only 8% of the available joint surface with a load of 32 lbs and increased to a maximum of only 15% even with a load of 118 lbs.. The total contact area (calculated as the average contact area/total available joint area) for each fossa was: STT 1.3%; SC 1.8%; CL 3.1%; TH 1.8%. The load distribution (calculated as the average contact area/total area of contact) through each joint was: STT = 17%; SC = 22%; CL = 39%; TH = 22% (Viegas, et al., 1990). This suggests that the primary load path in the wrist is through the capitate, with 50% of the load transmitted through the capitate to the scaphoid and lunate.

Posttraumatic Wrist

In addition to investigating the biomechanics of the normal wrist, researchers have been eager to evaluate the load characteristics in wrists with various types of post traumatic ligament instabilities, Kienböck's Disease, and various types of surgical treatments. A number of studies using pressure sensitive film have been performed in attempts to address these issues.

Radial Sided Perilunate Instability (Scapholunate Dissociation)

One type of ligament instability which has been simulated and studied is radial sided perilunate instability. Viegas et al. (1987) have studied the effects of progressive stages of radial sided perilunate instability using Mayfield's (1984) classification of stages of increasing instability. Wrists with increasing radial sided perilunate instability were shown to have areas of increased pressure in the scaphoid fossa and decreased pressure in the lunate fossa (Figure 7a and Figure 7b). The location within the wrist where these increased pressures developed correlated closely with areas in which degenerative changes develop in patients with the same type of radial sided perilunate instability (i.e. scaphoid instability) (Ruby, et al., 1988) (Figure 7c).

In all stages of perilunate instability, the contact areas remained a relatively small part of the overall joint surface. Pressures were significantly increased in wrists with stage III instability compared with normal wrists. The distance between the scaphoid and lunate contact areas changed little except when the wrist was placed in 20° of extension, neutral radioulnar deviation, and 90° of supination. In this position, there was a significant increase in the intercentroid distance. This correlates well with the fact that the stress view which best demonstrates the scapholunate gap on x-ray is this same position (i.e. AP, supinated, clenched fist view).

The effects of increasing perilunate instability on the load transfer characteristics of the wrist included a significant dorsal ulnar shift of the scaphoid centroid with increasing perilunate instability together with a less dramatic palmar ulnar shift of the lunate centroid (Viegas, et al., 1987). These changes in the centroids also correlate nicely with the changes in the carpal alignment with radial sided perilunate instability which has been called dorsal intercalated segment instability (DISI) (Linscheid, et al., 1972; Dobyns, et al., 1975).

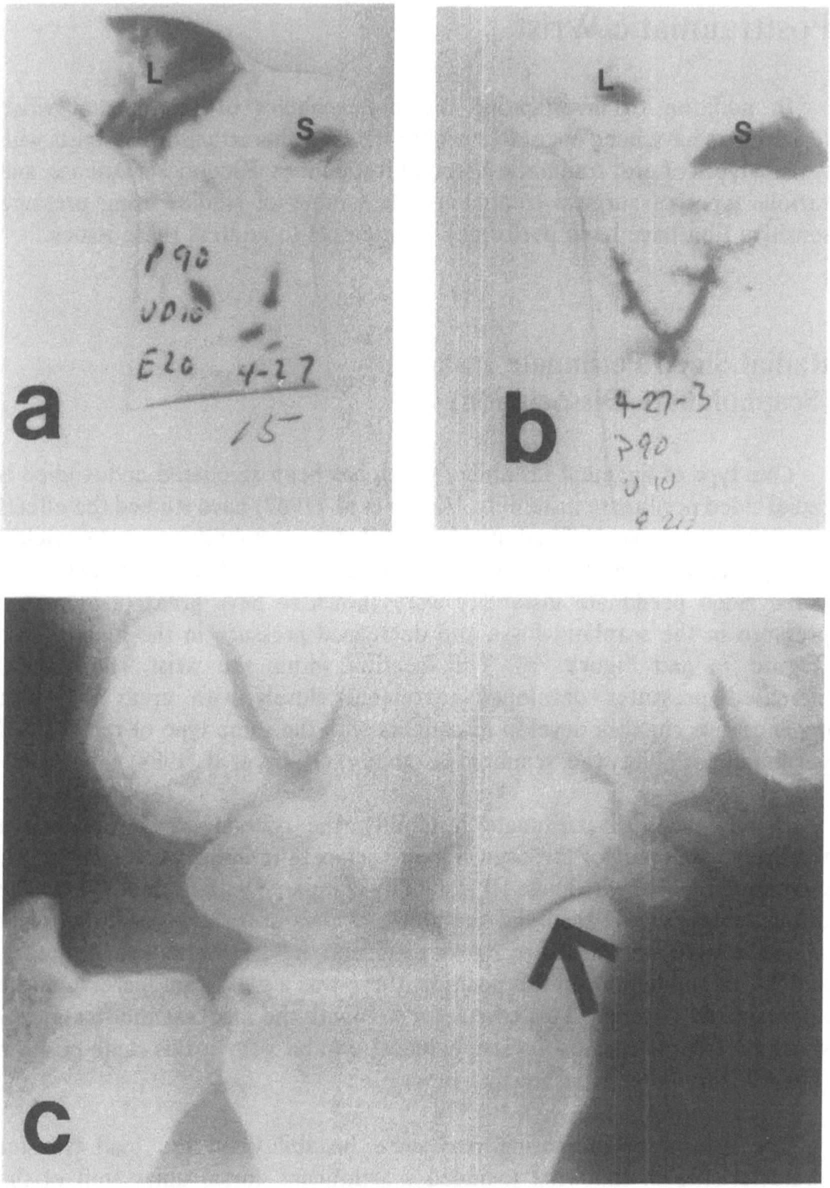

FIGURE 7: Top. Changes in contact area of a left wrist in the same position (90° pronation, 10° ulnar deviation, and 20° extension) at different stages (a) normal, (b) stage III. [Lunate contact area (L). Scaphoid contact area (S)]. (c) Bottom. Plain film illustrating the degenerative changes in the radioscaphoid joint of a wrist with stage III instability. The most dramatic changes are in the area that coincides with the most dramatic centroid, peak pressure, and contact area changes.

Blevens et al. (1989) studied the radiocarpal joint with simulated scaphoid instability from different combinations of ligament disruption. They studied the normal wrist as well as wrists with a scapholunate interosseous ligament tear (SLIL), wrists with a scaphotrapezial ligament complex tear (STLC), and wrists with a palmar intracapsular radiocarpal ligament tear (PIRL). They determined that the SLIL is the principal structure responsible for the maintenance of proximal scaphoid stability and a tear in this ligament causes dorsal translation of the scaphoid contact area. They also noted that partial rather than total scaphoid instability produces the highest scaphoid pressures, and that the totally destabilized scaphoid transmits less force than one with some restraints, thus leaving the lunate to take the remaining load (Blevens, et al., 1989). These combinations of ligament disruption are different from those described by Mayfield (1984) and studied by Viegas et al. (1987).

Ulnar Sided Perilunate Instability (Lunotriquetral Dissociation)

There is a great deal of conflicting information regarding the description of carpal instabilities, particularly with respect to the ulnar aspect of the wrist. This, in part, may arise from the duplicity and lack of uniformity in the use of terms such as dynamic and static, and the various classifications such as Carpal Instability Dissociative (CID) and Carpal Instability Non-Dissociative (CIND), Volar Intercalated Segment Instability (VISI) and Dorsal Intercalated Segment Instability (DISI), as well as midcarpal instability, which are not always quoted or utilized within the defined parameters the original authors had intended (Gilula, 1989). Recent efforts have been made to further understand and categorize the biomechanics and etiology of ulnar sided perilunate instability resulting in a VISI pattern.

A staging system for ulnar sided perilunate instability was developed based on a series of cadaver dissections and load studies by Viegas et al. (1990). This staging system follows the sequence: Stage I) partial or complete disruption of the lunotriquetral interosseous ligament, without clinical and/or radiographic evidence of dynamic or static VISI deformity; Stage II) complete disruption of the lunotriquetral interosseous ligament and disruption of the palmar lunotriquetral ligament, with clinical and/or radiographic evidence of dynamic VISI deformity; and Stage III) complete disruption of the lunotriquetral interosseous and the palmar lunotriquetral ligaments, attenuation or disruption of the dorsal radiocarpal ligament, with clinical and/or radiographic evidence of static VISI deformity (Figure 8).

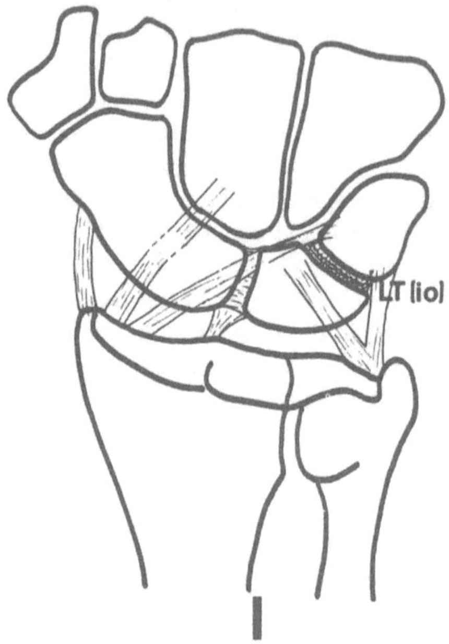

FIGURE 8a. Stage I ulnar-sided perilunate instability. A diagram depicting the disruption of the lunotriquetral interosseous ligament (LT[io]) in a stage I ulnar-sided perilunate instability.

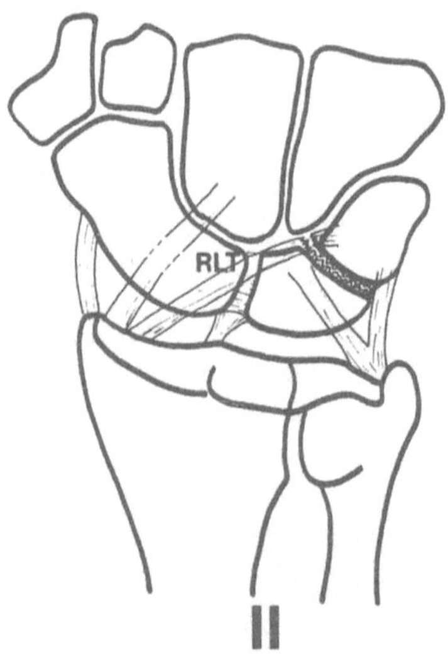

FIGURE 8b. Stage II ulnar-sided perilunate instability. A diagram depicting the additional disruption of the palmar radiolunotriquetral ligament (RLT), between the lunate and the triquetrum, in a stage II ulnar-sided perilunate instability.

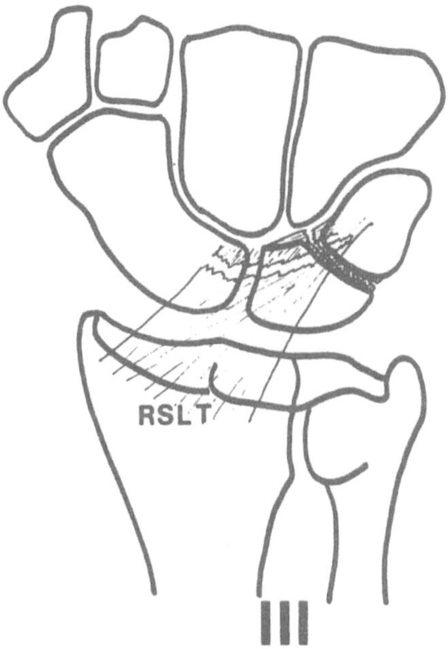

FIGURE 8c. Stage III ulnar-sided perilunate instability. A diagram depicting the additional disruption of the dorsal radioscapholunotriquetral ligament (RSLT), also called the dorsal radiocarpal ligament, in a stage III ulnar-sided perilunate instability.

In the cadaver studies, increased motion developed between the lunate and the triquetrum with a tear of the lunotriquetral interosseous ligament. However, with a disruption of this ligament alone, an appreciable dynamic or static VISI deformity was not evident and could not be induced. This is correlated with the findings of pressure studies, which overall, did not demonstrate significant differences in pressure distribution between the scaphoid and lunate fossa in the normal and a stage I instability. This would imply that the pressure distribution is not appreciably altered in cases where the lunotriquetral joint alone was disrupted (Viegas, et al., 1990). This is consistent with clinical findings that patients with incongruity between the lunate and the triquetrum generally had satisfactory clinical results (Minami, et al., 1986).

In a stage II ulnar-sided perilunate instability, a definite VISI deformity was evident during the application of a translational force at the dorsal aspect of the capitate and/or hamate, when the wrist was in neutral or some degree of flexion and radial, neutral or limited ulnar deviation. Sectioning the dorsal radiocarpal ligament and capsule at their attachment to the scaphoid and lunate, resulted in a stage III ulnar-sided perilunate instability and allowed a static VISI deformity to arise in the wrist. This static VISI deformity could only be demonstrated, however, in those same positions in which the dynamic VISI deformity could be attained. Whenever the wrist was brought into greater than 25° of ulnar deviation or into extension, this would result in changing the VISI position to a position of normal carpal alignment (Viegas, et al., 1990). Reduction of the VISI deformity by ulnar deviation of the wrist has been described clinically. The symptomatic "clunk" that patients often describe can often be reproduced by this maneuver (Lichtman, et al., 1981). This is still consistent with the classic definition of a VISI deformity since it applies to the capitolunate angle in the lateral radiograph with the wrist in neutral flexion and neutral radioulnar deviation. The clunk appears to occur when the head of the capitate suddenly shifts into the distal concavity of the lunate from its eccentrically loaded position on the volar lip of the lunate. The lunate is excessively flexed in VISI and is forced out of flexion by tension from the radioscaphocapitate ligament and scapholunate interosseous ligament, resulting from extension of the scaphoid as the wrist is extended or ulnarly deviated (Figure 9).

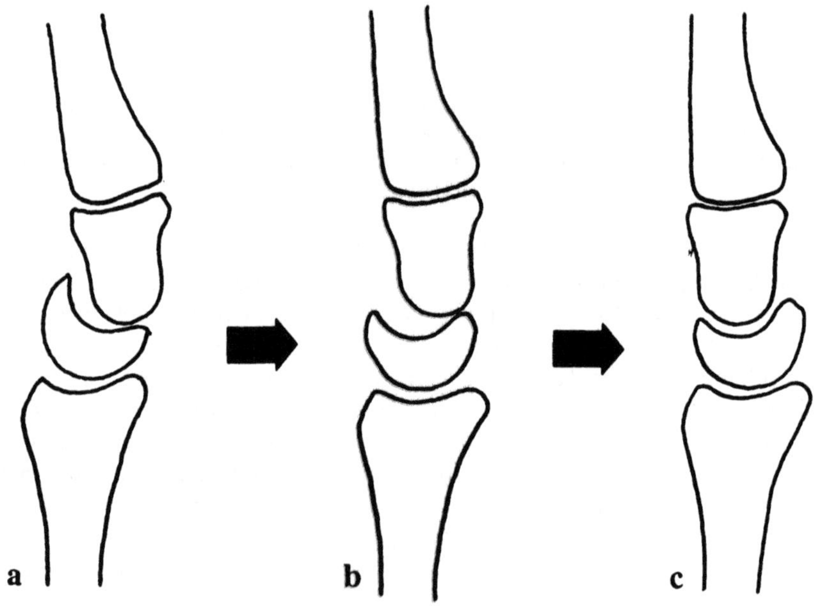

FIGURE 9: A series of diagrams demonstrating a) the excessively flexed posture of the lunate with the capitate eccentrically loaded on the volar lip of the lunate in a wrist with volar intercalated segment instability, then b) with extreme ulnar deviation and/or extension of the wrist the lunate is brought into a more neutral position by its scaphoid attachments allowing the capitate to transiently perch on the volar lip of the lunate until, c) it clunks back into the concavity of the lunate where it normally articulates.

The information from these pressure studies would suggest that the overall distribution of pressure centroids of the scaphoid and the lunate remains essentially the same throughout the progressive stages of ulnar-sided perilunate instability. The fact that even in stage III ulnar-sided perilunate instability, the high pressure centroids shifted to normal locations within portions of the functional range of motion of the wrist, would suggest that significant changes in the wear pattern in the radiocarpal joint would not change. This expectation, coincides with the clinical observation of sparing of the radiocarpal joint in patients with VISI deformities.

Fractures/Malunions/Non-unions

The investigation of the biomechanics of fractures and malunions is another area of interest that has been addressed by researchers utilizing Fuji prescale film. Radius, scaphoid, and ulna styloid fractures have been assessed using the normal load patterns and pressures in the wrist joint.

Distal Radius Fracture/Malunion

Distal radius fractures are the subject of much discussion (Jupiter and Masem, 1988; Bacorn and Kurtzke, 1953; Older, et al., 1965; Gartland and Werley, 1951; McQueen and Caspers, 1988; Short, et al., 1987; Solgaard 1984; Solgaard 1988; Villar, et al., 1987; Fernandez 1988; Lindstrom 1959; Stewart, et al., 1985; Ambrose and Posner, 1988; Cooney, et al., 1980; Rubinovich and Rennie, 1983; Taleisnik and Watson, 1984; Fernandez 1982; Colles 1814; Jenkins and Mintowt-Czyz, 1988; Overgaard and Solgaard, 1989). They are relatively common fractures and have been reported to have a high complication rate (Jupiter and Masem, 1988; Bacorn and Kurtzke, 1953; Ambrose and Posner, 1988; Cooney, et al., 1980; Rubinovich and Rennie, 1983; Taleisnik and Watson, 1984). Specific guidelines with respect to how much deformity can or should be accepted in treating distal radius fractures is lacking in the literature. This appears to be at least in part due to the difficulty in studying clinical outcomes of distal radius fractures where so many variables are involved.

Short et al. (1987) investigated the load characteristics in the wrist joint with simulated radius fractures. Wrists were tested in the normal condition and with 0° - 55° of radial angulation. It was determined that the ulnar force increased from 21% to 65% with increasing angulation. Additionally, with increased angulation, the radiolunate joint contact area shifted palmarly before disappearing. There was increased concentration of pressure in the radioscaphoid joint with the centroid migrating dorsally and radially and the ulnocarpal joint showed a shift in the centroid dorsally. At 40° angulation, the majority of the load is borne by the ulna and the pressure is concentrated at the dorsal radioscaphoid and ulnocarpal joints with no load in the radiolunate joint (Short, et al., 1987).

Pogue et al. (1990) tested the various components of simulated radius malunions. The components studied were radial shortening, radial inclination in the radioulnar plane and radial tilt in the dorso-palmar plane. In these simulated radius malunions studied, a decrease in the scaphoid contact area and an increase in the lunate contact area were seen after decreasing the radial inclination from normal (19° to 30°) to 10° - a change of only 9° to 20°. Variations in palmar inclination cause more concentrated contact areas, i.e. increased scaphoid and lunate high pressure contact areas without an appreciable change in the scaphoid/lunate load distribution, after angulating from normal (4° to 8° palmar) to 30° dorsal inclination - a change of 22° to 38°. While mild degrees of radial shortening caused a slight increase in the lunate pressure, more extreme shortening of 6 mm - 8 mm was noted to result in gross extra-articular ulnar impingement on the carpus.

This work supports the following conclusion and implies clinical guidelines for the reduction of fracture deformities and malunions of the radius. Distal radius fracture deformity or malunion with less than 2 mm of radial shortening, maintenance of at least 10° of radial inclination on the posteroanterior view and a change of less than 20° of palmar inclination on the lateral view, should not induce significant load changes compared to a normal wrist.

In addition, the earlier work on normal wrist load distribution would suggest preferred positions of immobilization for different fractures. For example, in a wrist with a distal radius volar lip fracture, placement of the wrist in extension should unload that fracture fragment (Pogue, et al., 1990).

Scaphoid Fracture/Non-union

Scaphoid fractures and non-union of the scaphoid continue to pose a clinical problem. Various studies have suggested that the natural history of a scaphoid non-union is the development of degenerative radiocarpal arthritis (Mack, et al., 1984; Ruby, et al., 1985; Vender, et al., 1987). Viegas, et al. (1991) addressed the question of how the load transfer characteristics of the wrist were altered by a fracture of the scaphoid proximal pole. Non-union and avascular necrosis of proximal pole scaphoid fractures continue to be a proportionally greater problem than other types of scaphoid fractures.

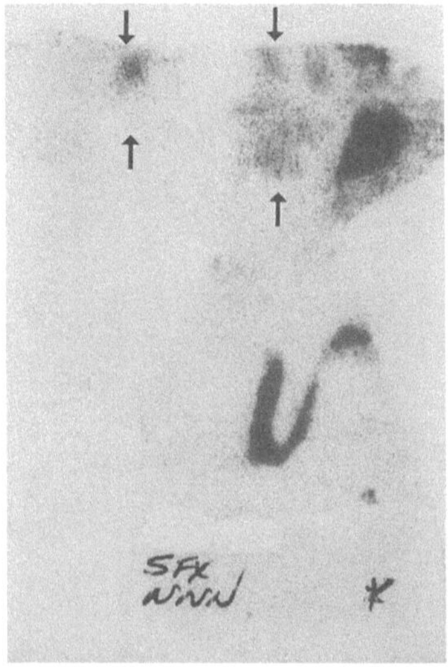

a)

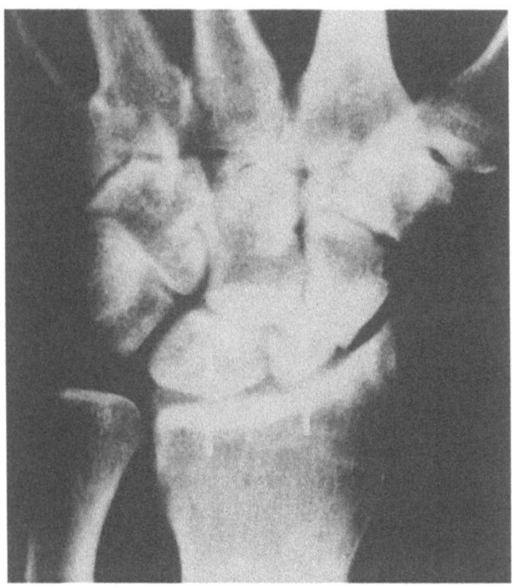

b)

FIGURE 10: A print of the pressure contact areas of the scaphoid and lunate prints of a right wrist in 20° of extension, neutral radioulnar deviation and neutral pronation/supination, oriented so that palmar is at the top and dorsal is at the bottom, a) showing scaphoid contact on the left and the lunate contact on the right with a scaphoid osteotomy demonstrating the type of decrease in load under the proximal pole and little or no decrease in load under the lunate (small arrows) and b) a radiograph of a wrist with a scaphoid non-union demonstrating the preservation of the joint space in these same areas (small arrows).

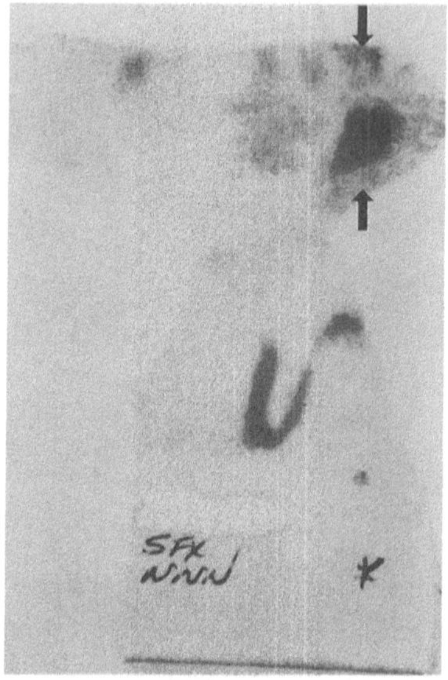

a)

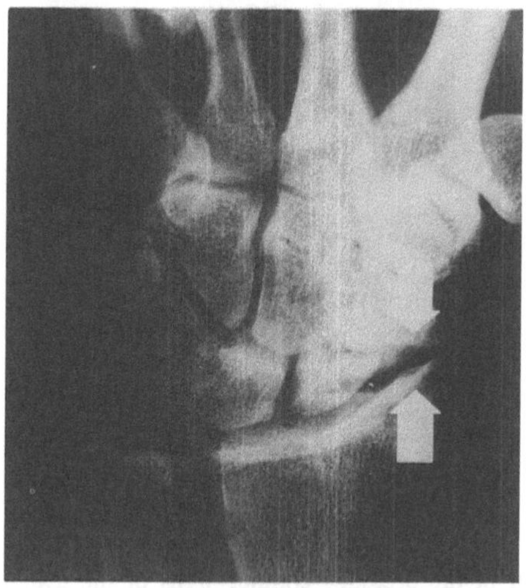

b)

FIGURE 11: A print of the pressure contact areas of the scaphoid and lunate prints of a right wrist, in 20° of extension, neutral radioulnar deviation and neutral pronation/supination, oriented so that palmar is at the top and dorsal is at the bottom, a) showing scaphoid contact on the left and the lunate contact on the right with a scaphoid osteotomy demonstrating the type of increase in load under the distal 2/3 of the scaphoid (large arrow) and b) a radiograph of a wrist with a scaphoid nonunion demonstrating the degenerative changes and loss of joint space in this same area (large arrow).

This work found that the amount of contact area born through the scaphoid fossa was essentially the same whether the scaphoid was intact, or a simulated proximal pole scaphoid fracture was made or the proximal pole scaphoid fragment was excised. The scaphoid contact area and pressure, although relatively constant, was redistributed following a simulated scaphoid proximal pole fracture, with increased contact area under the distal fragment and no change or a slight decrease in the contact area under the proximal fragment of the scaphoid. After resection of the proximal fragment, all scaphoid contact area and pressure was born by the distal scaphoid fragment. The contact area and pressure characteristics of the lunate remained unchanged in all conditions compared to normal. There were no significant changes in the locations of the centroids of the scaphoid segments and the lunate in any of the conditions tested.

These pressure studies demonstrate that a scaphoid osteotomy does not significantly affect the load distribution through the lunate and suggests that the load through the proximal scaphoid fragment is also essentially unchanged or slightly decreased. It does, however, acutely increase and concentrate the load distribution through the distal fragment of the scaphoid. These changes of increased load through the distal scaphoid fragment and decreased load through the proximal fragment were even more dramatic when the wrists were tested using greater force loads. The load born by the distal scaphoid fracture fragment may also increase over time in an in vivo setting with chronic compromise and attenuation of the associated ligamentous structures (Viegas, et al., 1991). The areas in which the distribution is not changed or is lessened coincide with areas that are observed clinically to be spared from degenerative changes even after three decades (Figure 10). The area in which the pressure distribution is increased, however, coincides with areas that are observed clinically to develop significant degenerative changes within one to two decades (Figure 11) (Viegas, et al., 1990; Mouchet and Belot, 1934; Gordon and Armstrong, 1968). This information would certainly question the reported treatment of scaphoid non-union by resection of the proximal pole of the scaphoid.

Distal Radioulnar Joint

Disruption of the distal radioulnar joint (DRUJ) alone or in association with distal radius fractures is another wrist problem that has been studied. Viegas et al. (1990) studied three stages of radioulnar instability: Stage 1) an avulsion fracture at the base of the ulna styloid; Stage 2) an avulsion fracture at the base of the ulna styloid plus disruption of the dorsal portion of the distal radioulnar joint capsule and; Stage 3) an avulsion fracture at the base of the ulna styloid, with disruption of the dorsal portion of the DRUJ capsule and disruption of the radioulnar interosseous membrane. All stages of radioulnar instability demonstrated a decrease in the lunate contact area in all wrist positions in which the forearm was in supination. In stage 3 instability, there was less lunate contact area in all wrist positions in which the forearm was in neutral pronation/supination. Also in stage 3 DRUJ instability, the lunate high pressure area centroid was abnormally palmar in all positions and the scaphoid high pressure area centroid was abnormally palmar in all wrist positions in which the forearm was in pronation or supination. This would suggest that a position of neutral pronation/supination might be a preferred position for immobilization in acute stage 1 and 2 injuries and that immobilization alone in any position may not be adequate treatment for stage 3 injuries (Viegas, et al., 1990).

Additionally, work performed by Pogue et al. (1990) has shown that an isolated ulna styloid fracture increases only the scaphoid/lunate area ratio and does not change the positions of the scaphoid and lunate high pressure areas or the scaphoid and lunate total contact areas. The fact that ulna styloid fractures change the load distribution so little may explain why they are often clinically asymptomatic. In simulating the various patterns and degrees of deformity of the radius, it was impossible to obtain displacements of the distal radius involving shortening greater than 4 mm, or angulations greater than approximately 20° with the ulna styloid and TFCC intact. This suggests that distal radius fractures with this displacement or more, not displaying an ulna styloid fracture, probably have a TFCC disruption (Pogue, et al., 1990).

Surgical Procedures

Pressure sensitive film and the findings from previous studies also utilizing pressure sensitive film have been utilized to assess the biomechanical efficacy of various types of surgical treatment for different conditions and injuries.

Surgical Treatments for Carpal Instabilities

Together with an increasing awareness over the recent years of the clinical entities known as carpal instabilities (Mayfield 1984; Mouchet and Belot, 1934; Gordon and Armstrong, 1968; Howard, et al., 1974; Jackson and Protas, 1981; Morawa, et al., 1976; Palmer, et al., 1978; Russell 1949; Watson and Hempton, 1980; Vaughan-Jackson 1949; Linscheid, et al., 1972; Mayfield, et al., 1980), there has been a concern that carpal instabilities are a precursor to degenerative changes in the wrist joint (Viegas, et al., 1987; Watson and Brenner, 1985; Watson and Ballet, 1984; Watson and Ryu, 1986; Blatt 1987).

Surgical treatments have been developed which involve soft tissue repair or reconstruction as well as limited intercarpal arthrodeses. Various limited carpal fusions were simulated by Viegas et al. (1991) to attempt to assess the biomechanical efficacy of limited carpal fusions for the treatment of scapholunate dissociation. Scaphoid-trapezium-trapezoid (STT) and scaphoid-capitate (SC) fusions were found to transmit almost all load through the scaphoid fossa. Scaphoid-lunate (SL), scaphoid-lunate-capitate (SLC) and capitate-lunate (CL) fusions all distributed load more proportionately through both scaphoid and lunate fossae, although not in exactly the same way as seen in the normal wrist. The relative positions of the carpal bones within a limited carpal fusion was also found to affect the load distribution in the wrist.

The SL, SLC or CL fusions, with attention to the relative carpal alignment within the limited fusion, appeared to offer more promise for treatment of perilunate instability biomechanically than the STT or SC fusions. On the other hand the significant loading of the scaphoid and unloading of the lunate seen in the study casts serious questions regarding the biomechanical efficacy of the SC and even the currently popular STT fusion, as treatment options for scapholunate instability (Viegas, et al., 1990).

Surgical Treatment for Kienböck's Disease

Lunatomalacia was first described by Kienböck (1910) and is characterized by necrosis of the lunate with secondary osteolytic and osteoblastic changes of collapse and repair. Ulnar negative variance has had a high correlation with Kienböck's disease and its role in the etiology of this condition has been questioned.

Various procedures used to treat Kienböck's disease have been simulated by Palmer et al. (1984), including increasing or decreasing the height of the ulna, closing and opening wedge osteotomies of the radius, and different types of limited carpal fusions. An attempt to evaluate whether an increase or decrease in ulnar height would affect the load transferred through the lunate was made (Palmer, et al., 1984). Normal wrists were tested in neutral extension/flexion, radial/ulnar deviation and rotation, wrists shortened or lengthened 2.5 mm, with and without the triangular fibrocartilage complex excised. Overall, an ulnar lengthening of 2.5 mm increased ulna pressure, increased radioscaphoid pressure and decreased pressure in the radiolunate fossa. An ulna shortening of 2.5 mm decreased ulna pressure, increased the radiolunate pressure and increased radioscaphoid pressure. Removal of the TFCC causes a shift of pressure centrally to the radiolunate articulation, thus decreasing the ulnolunate and the radioscaphoid articulation from normal loading (Palmer, et al., 1984).

Werner et al. (1989) investigated the effect of ulnar lengthening or radial shortening on the pressures at the distal radial ulnar joint. Wrists were tested in their normal condition as well as with combinations of radial shortening of 2.5 mm and 5.0 mm, radial displacement of 0 mm, 2 mm, and 4 mm, and an ulnar lengthening of 2.5 mm, 3.0 mm. They determined that the load distribution was changed comparably whether the ulna was lengthened or the radius was shortened. The distal radial ulnar joint pressure significantly increased from normal with a decrease in radial length of 2.5 mm and also increased with a combination of radial shortening of 2.5 mm and 2 mm radial displacement. The pressure decreased when the radius was shortened 2.5 mm and displaced 4 mm. When the radius was shortened 5.0 mm, the pressure increased significantly. It also increased if there was a corresponding radial displacement of 2 mm or 4 mm. With either ulnar lengthening or radial shortening, the centers of pressure were located more distally in the sigmoid notch (Werner, et al., 1989).

Joint leveling procedures have been shown to be useful in the treatment of Kienböck's disease in patients with negative ulnar variance. Some patients with Kienböck's disease, however, have positive ulnar variance, which contraindicates ulnar lengthening procedures. Thus, a study was undertaken by Werner et al. (1988) to ascertain the efficacy of a wedge osteotomy in unloading the radiolunate joint. Wrists were tested in the normal condition as well as with a radial opening osteotomy, a medial closing osteotomy, and a lateral closing osteotomy. The lateral opening and medial closing osteotomies both increased the force on the ulna. These procedures almost eliminated the radiolunate pressure while the ulnolunate pressure increased proportionate to the wedge angle. The lateral closing wedge osteotomy decreased force on the ulna. It also decreased the ulnolunate pressure, increased the radiolunate pressure, and shifted the pressure dorsally. Thus, biomechanically, the lateral opening or medial closing osteotomies appeared to be the best to unload the lunate in patient's with an ulnar neutral variance (Werner, et al., 1988).

In clinical situations where significant degenerative changes have occurred in limited portions of the wrist joint, one treatment alternative has been excision of one or more carpal bones and insertion of a silastic implant. Over the years since silastic carpal implants first became available, recommendations for their use and speculation of their load bearing characteristics, alone and with associated limited carpal fusions have been proposed (Watson and Ballet, 1984). With the increased concerns over silicon synovitis, a trend has evolved where a smaller implant is used, often with a limited carpal fusion, under the assumption that these modifications will shield the implant from bearing load.

In order to understand how a scaphoid implant acts biomechanically by itself and in associated with a limited carpal fusion, Viegas et al. (1991) studied the silastic scaphoid. It was demonstrated that the scaphoid silastic (high performance) implant, used as it is recommended, registers significant but diminished contact area and pressures on the radius than the normal scaphoid. The scaphoid silastic implant also had different pressure centroids than the normal scaphoid it replaces. Down-sizing the scaphoid implant decreased the load transmitted by the implant. The decreased contact area/pressure (i.e. load) through the scaphoid implant was compensated for an increase in the lunate load. The addition of a limited carpal fusion did not appear to appreciably decrease the load born by a scaphoid implant. Therefore, the silastic scaphoid implant is a load bearing implant even when undersized or placed in association with a limited carpal fusion (Viegas, et al., 1991). Admittedly the specimens studied in this experimental model did not have the associated DISI deformity often identified in progressive radial sided perilunate instability or the so called scapholunate advanced collapse (SLAC), wrist in which the use of a silastic scaphoid was recommended (Viegas, et al., 1991). In the cadaver models studied, the capsuloligamentous structures not directly attached to the scaphoid were protected and remained essentially intact. Despite this limitation, however, other studies would imply that with a progressive carpal collapse pattern, the scaphoid, and presumably a scaphoid implant, would sustain even greater load. With this in mind, a limited carpal fusion may serve the purpose of preventing late carpal collapse, if it has not yet developed and thereby avoid increasing loads on a scaphoid implant over time. It does not, however, appear to acutely diminish the load born by the implant and certainly does not completely unload the implant, as has been suggested.

In clinical situations where significant degenerative changes have occurred in the lunate, one treatment alternative has been excision of the lunate and insertion of a silastic lunate implant. Silastic prosthetic replacement of the lunate has been a common surgical procedure for the treatment of Kienböck's disease. However, these implants have been shown to deform and fragment over time.

Werner et al. (1987) measured the force and pressure distribution in the radial-ulnar carpal joint wrist which were intact, with the lunate excised and wrists with silastic lunate implants of various sizes. It was determined that the force through the ulna increased with lunate excision and with all lunate implants tested. The pressures at the radiolunate fossa that were eliminated with removal of the lunate were partially restored with the silastic replacements. They went on to perform two simulated limited carpal fusions (STT and SC) on these wrists in the normal condition, with the lunate excised and with the various implants. It was found that forces through the ulna decreased when the lunate was excised and either fusion was performed.

For each fusion, the ulnar force was increased compared to the normal and excised conditions when an implant the correct size or larger was in place, and decreased when a smaller implant was in place. These fusions also greatly reduced the pressure in the radiolunate fossa in the case of the smaller lunate implant. With either the correct size or larger size implant, pressure was restored to the radiolunate fossa despite these fusions.

If the lunate is excised, a subluxation of the scaphoid occurs and the triquetrum contacts the ulna. Although not statistically significant, the STT fusion was slightly more effective than the SC fusion in decreasing both the load through the ulna and the pressures at the radiolunate fossa. The condition that best simulated the normal forces through the ulna was a large lunate implant with either an STT or SC fusion. However, the contact areas were on the rim and not in the central region of the fossae. Thus, these experimental findings would suggest the use of carpal bone fusions in conjunction with an undersized lunate implant may decrease the loads placed on the silastic replacement which may minimize implant degradation (Werner, et al., 1987).

Summary

Research on the biomechanics of the human wrist using pressure sensitive film has yielded a substantial amount of information. This information has increased our basic knowledge of carpal morphology and mechanics. It has also enhanced our understanding of the development of degenerative arthritis in certain types of clinical conditions. This additional information has already resulted in a better understanding of some of the current treatment methods and has raised questions regarding some others. Ideally, the information which has been gathered will lead to more efficacious forms of treatment for various types of clinical problems in the wrist. This research into wrist biomechanics is by no means a completed work; but rather, the foundation upon which continuing studies will be based.

References

Ambrose L, Posner MA: Biplanar Osteotomy for the Treatment of Malunited Colles' Fractures. Presented at The 43rd Annual Meeting of the American Society for Surgery of the Hand. Baltimore, Maryland. Sept. 14-17, 1988.

Bacorn RW, Kurtzke JF: Colles Fracture. A Study of Two Thousand Cases from the New York State Workman's Compensation Board. Journal of Bone and Joint Surgery 1953;35:643-658.

Berger RA, Crowninshield RD, Flatt AE: The three-dimensional rotational behaviors of the carpal bones. Clin. Orthop., 1982;167:303-310.

Blatt G: Capsulodesis in Reconstructive Hand Surgery: Dorsal Capsulodesis for the Unstableet Scaphoid and Volar Capsulodesis Following Excision of the Distal Ulna. Hand Clinics 1987;3:81-102.

Blevens AD, Light TR, Jablonsky WS, Smith DG, Patwardhan AG, Guay ME, Woo TS: Radiocarpal Articular Contact Characteristics with Scaphoid Instability. Journal of Hand Surgery 1989;14A:781-790.

Colles A: On the Fractures of the Carpal Extremity of the Radius. Edinb. Medical Surgery Journal 1814;10:182-186.

Cooney WP, Dobyns JH, Linscheid RL: Complications of Colles' Fracture. Journal of Bone and Joint Surgery 1980;62:613-619.

Dobyns JH, Linscheid RL, Chao EYS, Weber ER, Swanson GE: Traumatic Instability of the Wrist. Instructional Course Lectures, AAOS, St. Louis, Missouri. The C.V. Mosby Co. 1975;pp.182-199.

Fernandez DL: Radial Osteotomy and Bowers Arthroplasty for Malunited Fractures of the Distal end of the Radius. Journal of Bone and Joint Surgery 1988;70:1538-1551.

Fernandez DL: Correction of Post-Traumatic Wrist Deformity in Adults by Osteotomy, Bone-Grafting, and Internal Fixation. Journal of Bone and Joint Surgery 1982;64:1164-1178.

Gartland JJ Jr, Werley CW: Evaluation of Healed Colles' Fractures. Journal of Bone and Joint Surgery 1951;33:895-907.

Gilula LA: Ligament Instability of the Wrist: Discussion of Current Classification Systems. Wrist Investigators' Workshop. Paris, France. April 5, 1989.

Gordon WD, Armstrong MD: Rotational Subluxation of the Scaphoid. The Canadian Journal of Surgery 1968;11:306-314.

Howard FM, Fahey T, Wojcik E: Rotatory Subluxation of the Navicular. Clinical Orthopaedics and Related Research 1974;104:134-139.

Ishizuka, Takao: "Prescale" Pressure Measuring Sheet, Fuji Photo Film Co., Ltd.

Jackson WT, Protas JM: Snapping Scapholunate Subluxation. The Journal of Hand Surgery 1981;6:590-594.

Jenkins NH, Mintowt-Czyz WJ: Mal-Union and Dysfunction in Colles' Fracture. Journal of Hand Surgery 1988;13:291-293.

Jupiter JB, Masem M: Reconstruction of Post-Traumatic Deformity of the Distal Radius and Ulna. Hand Clinics 1988;4:377-390.

Kienböck R: Uber traumatishe Malazie des Mondbeins, und ihre Folgezustande: Entartungsformen und Kompressions Frakturen. Fortschr Roengenstr 1910;16:77.

Lange A de, Kauer JMG, Huiskes R: Kinematic behavior of the human wrist joint: a roentgen-stereophotogrammetric analysis. J. Orthop. Res., 1985;3:56-64.

Lichtman DM, Schneider JR, Swafford AR, Mack GR: Ulnar Midcarpal Instability-Clinical and Laboratory Analysis. Journal of Hand Surgery. 1981;6:515-523.

Lindstrom A: Fractures of the Distal end of the Radius. Acta Orthop. Scand. Suppl. 41, 1959.

Linscheid RL, Dobyns JH, Beabout JW, Bryan RS: Traumatic Instability of the Wrist. The Journal of Bone and Joint Surgery 1972;54A:1612-1632.

Mack GR, Bosse MJ, Gelbermann RH, Yu E: The Natural History of Scaphoid Non-Union. Journal of Bone and Joint Surgery 1984;66A:504-509.

Mayfield JK: Patterns of Injury to Carpal Ligaments. Clinical Orthopaedics and Related Research 1984;187:36-42.

Mayfield JK, Johnson RP, Kilcoyne RK: Carpal Dislocations: Pathomechanics and Progressive Perilunar Instability. The Journal of Hand Surgery 1980;5:226-241.

McQueen M, Caspers J: Colles Fracture: Does the Anatomical Result Affect the Final Function? Journal of Bone and Joint Surgery 1988;70:649-651.

Minami A, Ogino T, Ohshio I, Minami M: Correlation Between Clinical Results and Carpal Instabilities in Patients after Reduction of Lunate and Perilunar Dislocations. Journal of Hand Surgery. 1986;11B:213-220.

Morawa LG, Ross PM, Schock CC: Fractures and Dislocations Involving the Navicular-Lunate Axis. Clinical Orthopaedics and Related Research 1976;118:48-53.

Mouchet A, Belot J: Poignet a Ressaut (Subluxation Mediocarpienne en avant). Bull Mem Soc Nat Chir 1934;60:1243-1244.

Older TM, Stabler EV, Cassebaum WH: Colles Fracture: Evaluation and Selection of Therapy. Journal of Trauma 1965;5:469-476.

Overgaard S, Solgaard S: Osteoarthritis After Colles' Fracture. Orthopedics 1989;12:413-416.

Palmer AK, Werner FW: Biomechanics of the Distal Radioulnar Joint. Clin Orthop 1984;187:26-35.

Palmer AK, Dobyns JH, Linscheid RL: Management of Post-traumatic Instability of the Wrist Secondary to Ligament Rupture. The Journal of Hand Surgery 1978;3:507-532.

Pogue DJ, Viegas SF, Patterson RM, Peterson PD, Jenkins DK, Sweo TD, Hokanson JA: The Effects of Distal Radius Fracture Malunion on Wrist Joint Mechanics. Journal of Hand Surgery 1990;15A:721-727.

Rieck B, Paar O, Bernett P: Inter-articular Pressure Measurement. A New Method for the use of Pressure Measuring Film 'Prescale'. Z Orthop., Nov-Dec. 1984;122(6):pp.841-842.

Rubinovich RM, Rennie WR: Colles' Fracture: End Results in Relation to Radiologic Parameters. Canadian Journal of Surgery 1983;26:361-363.

Ruby LK, Cooney III WP, An KN, Linscheid RL, Chao EYS: Relative motion of selected carpal bones: A kinematic analysis of the normal wrist. J. Hand Surg, 1988;13A:1-10.

Ruby LK, Stinson J, Belsky MR: The Natural History of Scaphoid Non-Union: A Review of Fifty-Five Cases. Journal of Bone and Joint Surgery 1985;67A:428-432.

Rudert MJ, et al.: Loading Characteristics of Presensor: A Pressure Sensitive Film, Orthopaedic Research Meeting, 1988.

Russell TB: Inter-Carpal Dislocations and Fracture-Dislocations, A review of fifty-nine cases. The Journal of Bone and Joint Surgery 1949;31B:524-531.

Short WH, Palmer AK, Werner FW, Murphy DJ: A Biomechanical Study of Distal Radius Fractures. Journal of Hand Surgery 1987;12:529-534.

Solgaard S: Function after Distal Radius Fracture. Acta Orthop Scand 1988;59:39-42.

Solgaard S: Classification of Distal Radius Fractures. Acta Orthop Scand 1984;56:249-252.

Stewart HD, Innes AR, Burke FD: Factors Affecting the Outcome of Colles' Fracture: an Anatomical and Functional Study. Injury 1985;16:289-295.

Taleisnik J, Watson HK: Midcarpal Instability Caused by Malunited Fractures of the Distal Radius. Journal of Hand Surgery 1984;9:350-357.

Vaughan-Jackson OJ: A Case of Recurrent Subluxation of the Carpal Scaphoid. The Journal of Bone and Joint Surgery 1949;31B:532-533.

Vender MI, Watson HK, Wiener BD, Black DM: Degenerative Change in Symptomatic Scaphoid Non-union. Journal of Hand Surgery 1987;12A:514-519.

Viegas SF, Patterson RM, Hillman GR, Peterson PD: The Simulated Scaphoid Proximal Pole Fracture: A Biomechanical Study. Journal of Hand Surgery. 1991;16A (in press).

Viegas SF, Patterson RM, Peterson PD, Crossley M, Foster R: The Silastic Scaphoid: A Biomechanical Study. Journal of Hand Surgery 1991;16A:91-97.

Viegas SF, Patterson RM, Peterson PD, Pogue DJ, Jenkins DK, Sweo TD, Hokanson JA: The Evaluation of the Biomechanical Efficacy of Limited Intercarpal Fusions for the Treatment of Scapholunate Dissociation. Journal of Hand Surgery 1990;15A:120-128.

Viegas SF, Patterson RM, Peterson PD, Pogue DJ, Jenkins DK, Sweo TD, Hokanson JA: Ulnar Sided Perilunate Instability: An Anatomic and Biomechanic Study. Journal of Hand Surgery 1990;15A:268-278.

Viegas SF, Patterson RM, Todd P, McCarty P: Load Transfer Characteristics of the Midcarpal Joint. Presented in part at the Wrist Biomechanics Symposium, Wrist Biomechanics Workshop, Mayo Clinic, Rochester, Minnesota, October 7, 1990.

Viegas SF, Pogue DJ, Patterson RM, Peterson PD: The Effects of Radioulnar Instability on the Wrist: A Biomechanical Study. Journal of Hand Surgery. 1990;15A:728-732.

Viegas SF, Patterson RM, Peterson PD, Roefs J, Tencer A, Choi S: The Effects of Various Load Paths and Different Loads on the Load Transfer Characteristics of the Wrist. Journal of Hand Surgery 1989;14A:458-465.

Viegas SF, Tencer AF, Cantrell J, Chang M, Clegg P, Hicks C, O'Meara C, Williamson JB: Load Transfer Characteristics of the Wrist: Part I. The Normal Joint. Journal of Hand Surgery 1987;12A:971-978.

Viegas SF, Tencer AF, Cantrell J, Chang M, Clegg P, Hicks C, O'Meara C, Williamson JB: Load Transfer Characteristics of the Wrist: Part II. Perilunate Instability. Journal of Hand Surgery 1987;12A:978-985.

Villar RN, Marsh D, Rushton N, Greatorex RA: Three Years after Colles' Fracture. Journal of Bone and Joint Surgery 1987;69:635-638.

Watson HK, Ryu J: Evolution of Arthritis of the Wrist. Clinical Orthopaedics and Related Research 1986;202:57-67.

Watson HK, Brenner LH: Degenerative Disorders of the Wrist. The Journal of Hand Surgery 1985;10A:1002-1006.

Watson HK, Ballet FL: The SLAC Wrist: Scapholunate Advanced Collapse Pattern of Degenerative Arthritis. The Journal of Hand Surgery 1984;9A:358-365.

Watson HK, Hempton RF: Limited Wrist Arthrodesis. I. The Triscaphoid Joint. The Journal of Hand Surgery 1980;5:320-327.

Werner FW, Murphy DJ, Palmer AK: Pressures in the Distal Radioulnar Joint: Effect of Surgical Procedures used for Kienböck's Disease. J. Orthopaedic Research, 1989;7(3).

Werner FW, Palmer AK, Utter RG: Distal Radial Osteotomy for the Treatment of Kienböck's Disease: A Biomechanical Study, 34th Annual Meeting, Orthopaedic Research Society, February 1-4, 1988.
Werner FW, Murphy DJ, Palmer AK: Silastic Synovitis Following Lunate Replacement A Biomechanical Evaluation of Preventative Measures, 33rd Annual Meeting, Orthopaedic Research Society, January 19-22, 1987.

Chapter 7

Strain Gauge Measurement in Carpal Bone

V.R. Masear

Introduction

Peste (Nahigian, et al., 1970) described avascular necrosis of the lunate in 1843, but Kienböck (1910) provided the first complete description of the condition which has since borne his name. Over the past 60 years numerous treatment regimens have arisen (Table 1). Far from being resolved, the treatment controversy seems to be increasing.

Hulten (1928) implicated ulnar shortening relative to the radius (negative ulnar variance) as an important etiological factor in avascular necrosis of the lunate. Subsequent reports have supported the increased incidence of lunate avascular necrosis in ulnar negative wrists. (Beckenbaugh, et al., 1980; Chan and Huang, 1971; Gelberman, et al., 1975). Persson (1945) postulated that the lunate was unevenly loaded between the unyielding lunate fossa of the radius and the compressible triangular fibrocartilage complex. Rossak (1967) felt the lunate would be subject to its greatest shear stress in forced ulnar deviation. An ulnar negative variant would increase lunate shear and compressive stress. Relative ulnar lengthening would thus decrease the shear stresses on the lunate and probably also buttress the ulnar column.

Chuinard (1985) advocated capitohamate (CH) arthrodesis as a method of preventing further collapse by removing the compressive effect of the capitate on the lunate. Watson (1985) recommended scaphotrapeziotrapezoidel (STT) arthrodesis to both remove the compressive stress from the lunate and treat the rotary subluxation of the scaphoid that accompanies Kienböck's disease with collapse.

TABLE 1. Treatment Methods Used in Kienböck's Disease.

Radial shortening	(Hulten 1928; Eiken and Niechajev, 1980; Almquist 1982; Moberg 1985)
Ulnar lengthening	(Persson 1945; Tillery 1968; Armistead, et al.; Sundberg and Linscheid, 1984)
Immobilization	(Cave 1939; Stahl; Dornan 1949; Tajima 1966)
Lunate excision	(Cave 1939; Dornan 1949; Tajima 1966; Gillespie 1961; McMurtry, et al., 1978; Youm and Flatt, 1980)
Lunate drilling	(Tajima 1966)
Bone grafting lunate	(Tajima 1966)
Carpal tunnel decompression	(Codega, et al., 1973)
Lunate excision and soft tissue interposition	(Nahigian, et al., 1970; Kato, et al., 1986)
Wrist arthrodesis	(Dornan 1949; Lichtman, et al., 1977)
Limited carpal fusion	(Graner, et al., 1966; Watson, et al., 1985; Chuinard 1985)
Proximal row carpectomy	(Lichtman, et al., 1977)
Vascular pedicle grafting	(Hori, et al., 1979; Lunn and Ashwell, 1985)
Lunate replacement arthroplasty	(Evans, et al., 1986; McMurtry, et al., 1978; Youm and Flatt, 1980; Lichtman, et al., 1977; Lippman and McDermott, 1949; Danis 1951; Agerholm 1963; Swanson 1970; Schuh, et al., 1975; Roca, et al., 1976; Beckenbaugh, et al., 1980; Stark, et al., 1981; Ramakrishna, et al., 1982)

With the exception of nonoperative methods, most treatments for Kienböck's disease have demonstrated some success. However, the joint leveling procedures have perhaps exhibited the most consistent favorable outcomes. Immobilization or no treatment often lead to persistent pain with increasing collapse and joint degeneration (Persson 1945; Tillery 1968; Stahl; Evans, et al., 1986; Lichtman, et al., 1977; Beckenbaugh, et al., 1980). Lunate excision and prosthetic replacement have led to carpal shift and progressive collapse (Evans, et al., 1986; Gillespie 1961; McMurtry, et al., 1978; Youm and Flatt,1980; Agerholm 1963; Stark, et al., 1981). Cyst formation, ulnar styloid changes, pain with ulnar deviation and dorsiflexion, and limitation of ulnar deviation secondary to abnormal contact between the ulnar styloid and triquetrum have followed lunate excision (Tillery 1968; Gillespie 1961). Besides progressive collapse, replacement arthroplasty of the lunate has been associated with dislocations, median nerve paresthesias, degenerative changes, bony cysts, and decreased grip strength (Evans, et al., 1986; McMurtry, et al., 1978; Kato, et al., 1986; Roca, et al., 1976; Beckenbaugh, et al., 1980; Stark, et al., 1981; Ramakrishna, et al., 1982; Michon 1985). Progressive collapse of the lunate and radiological deterioration have usually not been seen after the joint leveling procedures (Eiken and Niechajev, 1980; Almquist 1982; Moberg 1985; Tillery 1968; Armistead, et al.; Sundberg and Linscheid, 1984). Wrist arthrodesis and proximal row carpectomy are salvage procedures and should not be used in the early stages of Kienböck's disease before there is joint deterioration. Proximal row carpectomy is often functionally unsatisfactory leading to diminished grip strength and limited motion (Michon 1985). Popularized by Watson et al. (1985), recent treatment has shifted toward STT arthrodesis. However, complications following STT fusion have been significant. These have included nonunion, pain with motion, reflex sympathetic dystrophy, infection, and neuromas (Kraemer, et al., 1988; Bax, et al., 1988). Watson and Rogers (1988) have reported painful radial styloid impingement in 33% within two years after surgery and now recommend radial styloidectomy in conjunction with STT arthrodesis.

Biomechanical studies of the wrist have been constructed in an attempt to resolve the dilemma of the best treatment for Kienböck's disease, but have instead added to the dispute. Trumble et al. (1986) found joint leveling procedures (radial shortening and ulnar lengthening) and STT fusions beneficial in relieving lunate load. CH fusion was ineffective. Palmer and Werner (1982) demonstrated unloading of the lunate fossa by load transference to the scaphoid fossa with both STT and scaphocapitate (SC) fusions. Radial shortening with or without radial displacement, ulnar lengthening, lateral opening wedge radial osteotomy, and medial closing wedge radial osteotomy all consistently unloaded the lunate fossa of the radius by redistributing load to both the ulnar column and scaphoid fossa. CH fusion and carpal tunnel release did not relieve load from the lunate fossa. In a similar study Garcia-Elias (1988) found that STT and SC fusions did not relieve load from the lunate fossa nearly as well as did ulnar lengthening. The biomechanical investigation reported here was conducted to measure local strains occurring on the dorsal surface of the lunate.

Biomechanical Testing

Preserved cadaveric upper limbs were amputated through the distal humerus. Each wrist was x-rayed from posterior to anterior with the elbow flexed to 90° and the forearm in neutral rotation. The distance between a line marking the cortical bone of the ulnar head and one marking the distal edge of the sclerotic region of the radius was measured as the ulnar variance. This method was shown by Palmer et al. (1982) to be reliable and reproducible. Any specimen with notable arthritic changes or collapse was discarded.

Twenty limbs were stripped of soft tissues, leaving the skeletal and ligamentous structures intact. The fingers were disarticulated at the metacarpophalangeal joints and both the metacarpals and elbows were then potted in laboratory gypsum stone. The capsule was dissected off the dorsal lunate to expose the entire nonarticular surface. The bone surface was dried with alcohol and degreased with Freon TF. A thin coat of lacquer spray was applied to seal the pores in the bone. The dried lacquer surface was abraded with 400 grit SiC paper. The surface was then cleaned with M Prep Conditioner A followed by M Prep Neutralizer. The contact surface of the strain gauge was then coated with a catalyst and bonded to the prepared areas with methyl-2-cyanoacrylate adhesive. Rectangular rosette strain gauges (series WA-13-060 WR-120, Micro Measurements Division, Measurement Group, Inc., Raleigh, N.C.) were utilized to completely define the strain field since the direction of the principal axes was unknown (Figure 1). The dorsal surface of the distal radius was similarly prepared and the gauge leads soldered to tabs attached to the radius. Slack was left in the leads as a method of providing strain relief. The solder tabs were in turn connected to the switch and balance unit with a quarter bridge circuit with 426-DFV wires. A strain indicator unit (P-3500, Instruments Division, Measurements Group, Raleigh, North Carolina) was used to digitally monitor strain.

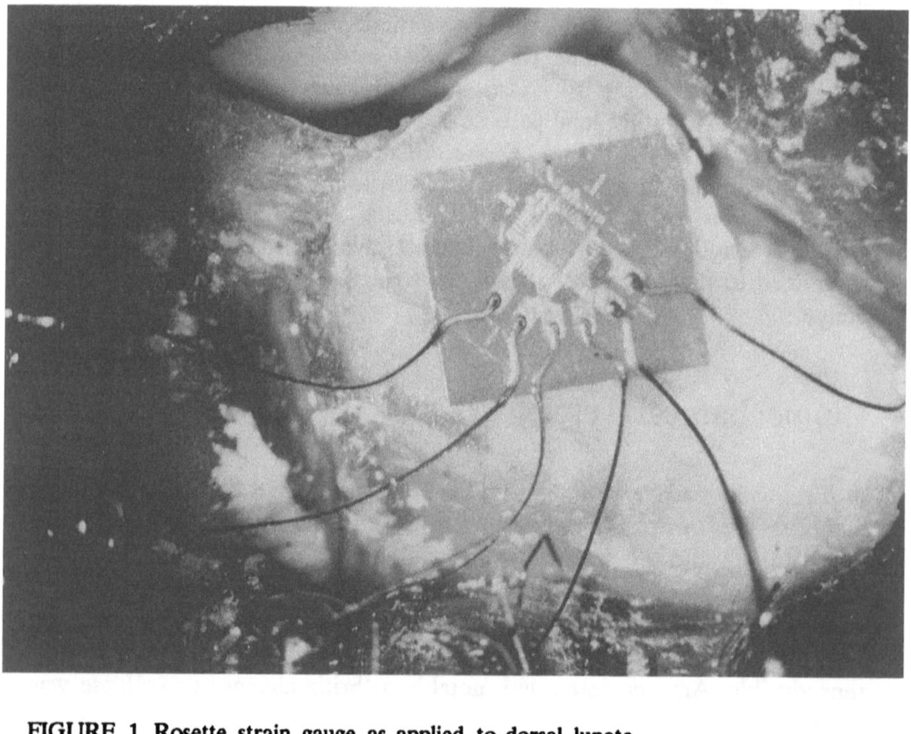

FIGURE 1. Rosette strain gauge as applied to dorsal lunate.

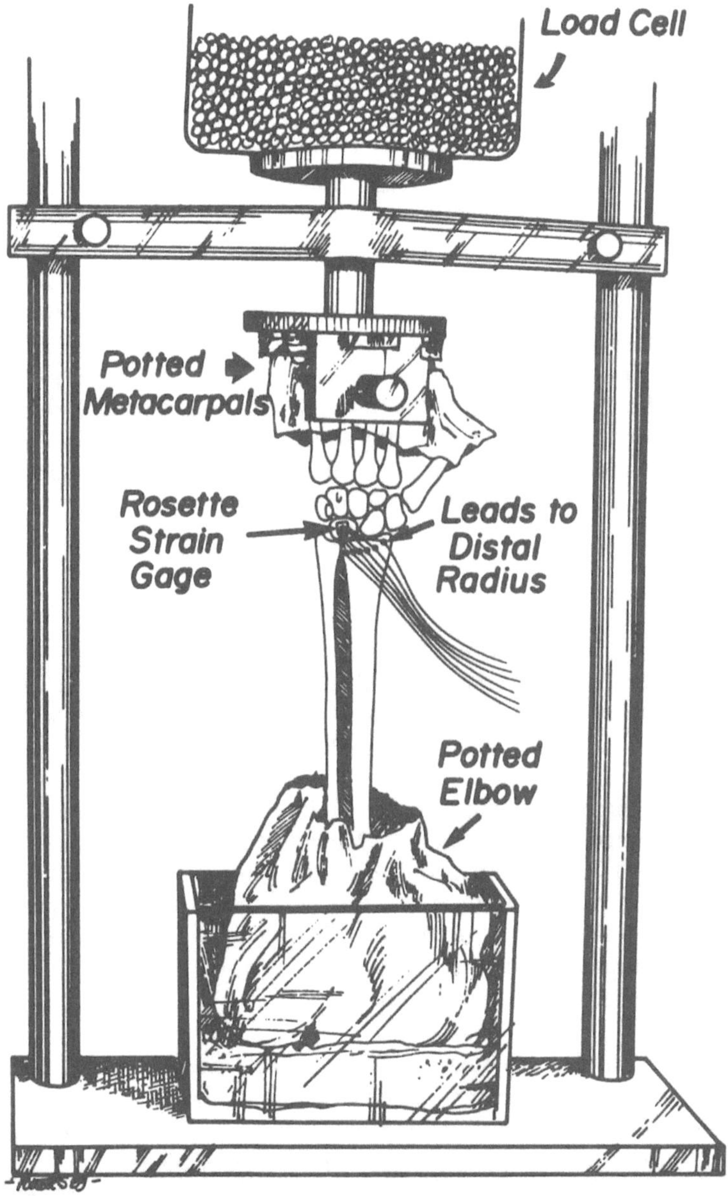

FIGURE 2. Artist's illustration of limb in load frame.

The limbs were then mounted in the load frame (Figure 2) by clamping the potted elbows and metacarpals. All testing was done with the wrists in neutral position. After zeroing of the strain gauge indicator, dead weights from zero to 10 Kg were applied in 2.5 Kg increments. Each 2.5 Kg load was allowed to act for two minutes before strain readings were taken. After the full 10 Kg load had been applied and removed a final 0 Kg load strain reading was recorded. This entire incremental loading procedure was performed and recorded three times in each specimen.

Following the initial testing each limb was randomized to receive a simulated CH or STT arthrodesis. Nine STT and 11 CH fusions were performed. Small neutralization plates, screws, and Kirschner wires were used for fixation of the arthrodeses. After fixation each specimen was subjected to the same method of loading and strain readings that had been done prior to the simulated fusions. In addition, the four limbs which demonstrated negative ulnar variance had quadrilateral external fixators applied to their ulnas. The bone was then marked prior to osteotomy of the ulna and each ulna lengthened by 2, 3, and 4 mm. Strain measurements were taken following each lengthening. The negative ulnar variance in these limbs was 3 mm. Therefore, lengthening of 2 mm still left a negative variance and lengthening of 4 mm created a positive ulnar variance.

Results

Maximum principal strains and maximum shear strains were calculated. Each test value was compared with its own pre-fixation control using the general linear models procedure and analysis of variance technique.

Table 2 summarizes the class level information from the five load levels and the complete data from all tests. The number of observations refer to the number of strain readings obtained in the given case and include all the specimens on which the given test was performed. The levels provide the number of "values" in a given class. Values provide information on the specimens included, the type of test performed, the load levels considered, and the number of runs performed.

Table 3 provides a summary of the statistical significance obtained for the two variables, the maximum principal strain (EMAX) and the maximum shear strain (2R) of the five load levels and complete data for all tests. Significance values for the four classes from each of the tests are given. The significance values of the test type are identified with an S (significant) or NS (not significant). Identification with a "+" (positive) sign refers to a test which increased the strain under consideration and "-" (negative) sign to a test which decreased the strain under consideration. This was determined from the EMAX and 2R values of the tests and the controls derived from the rosette strain data.

TABLE 2. General Linear Models Procedure:
Class Level Information For Five Load Levels, Complete Data

TEST*	OBS.**	CLASS	LEVELS	VALUES
STT	270	SPECIMEN	9	1 11 13 16 19 20 24 29 30
		TYPE	2	0 1
		LEVEL	5	2.5 5 7.5 10 0
		RUN	3	1 2 3
CH	330	SPECIMEN	11	2 3 5 7 9 12 14 22 23 27 31
		TYPE	2	0 2
		LEVEL	5	2.5 5 7.5 10 0
		RUN	3	1 2 3
4 mm UL	30	TYPE	2	0 3
		LEVEL	5	2.5 5 7.5 10 0
		RUN	3	1 2 3
2 mm UL	30	TYPE	2	0 4
		LEVEL	5	2.5 5 7.5 10 0
		RUN	3	1 2 3
3 mm UL	90	SPECIMEN	3	15 18 28
		TYPE	2	0 5
		LEVEL	5	2.5 5 7.5 10 0
		RUN	3	1 2 3

* Test: STT — Triscaphe fusion
CH — Capitate-hamate fusion
4 mm UL — 4 mm ulnar lengthening
2 mm UL — 2 mm ulnar lengthening
3 mm UL — 3 mm ulnar lengthening

** OBS.: Number of strain observations per test

TABLE 3. General Linear Models Procedure: Significance Values for Five Load Levels, Complete Data

DEP. Var.	TEST*				
EMAX**					
Pr > F	STT	CH	4 mm UL	2 mm UL	3 mm UL
SPECIMEN	0.0001	0.0001	------	------	0.0001
TYPE	0.0002	0.1375	0.0004	0.0441	0.0001
	(S+)	(NS+)	(S+)	(S-)	(S-)
LEVEL	0.0001	0.0001	0.0022	0.2149	0.0001
RUN	0.0010	0.4493	0.2415	0.0010	0.0621

DEP. Var.	TEST				
2R					
Pr > F	STT	CH	4 mm UL	2 mm UL	3 mm UL
SPECIMEN	0.0001	0.0001	------	------	0.0001
TYPE	0.0001	0.0095	0.5341	0.0239	0.6068
	(S+)	(S+)	(NS+)	(S-)	(NS-)
LEVEL	0.0001	0.0001	0.1789	0.4591	0.0001
RUN	0.0294	0.8451	0.3953	0.0091	0.3170

* STT — Triscaphe fusion
CH — Capitate-hamate fusion
4 mm UL — 4 mm ulnar lengthening
2 mm UL — 2 mm ulnar lengthening
3 mm UL — 3 mm ulnar lengthening
DEP. Var. — Dependent variable

** EMAX — Maximum principal strain
Pr > F — Probability greater than F
2R — Maximum shear strain
S — Significant
NS — Not significant
\+ — Test increased strain
\- — Test decreased strain

STT fusion consistently and significantly increased both the principal (compressive) strain and shearstrain on the lunate. Ulnar lengthening of 2 mm gave variable and inconsistent results. Ulnar lengthening of 3 mm significantly reduced the maximum principal strain, and although the results did not reach significance shear strain was also reduced. Ulnar lengthening of 4 mm was statistically significant for increasing principal strain. Ulnar lengthening of 4 mm also consistently increased shear strain, but the values did not reach statistical significance. CH fusion showed a trend toward decreasing principal strain but increasing shear strain. Neither value for CH fusion attained statistical significance.

Correlation Of Biomechanical Results

Progressive lunate collapse in Kienböck's disease results from continued loading. If load can be relieved from the lunate, the bone should be able to revascularize and heal without further collapse. We have attempted to identify the best method of unloading the lunate. By placing strain gauges directly on the lunate we have measured forces in the bone itself. Measuring load on the lunate fossa of the radius with pressure sensitive film as reported in several studies (Palmer and Werner, 1988; Garcia-Elias 1988), does not depict the amount of strain actually incurred by the lunate.

The STT fusions in this study were performed with the wrists in neutral so the scaphoid would lie at a 45° angle to the long axis of the radius. This has been the recommended position clinically. Watson et al. (1985) feel overcorrection of the collapse problem by bringing the scaphoid more into alignment with the radius will significantly limit motion and result in joint incongruity in the scaphoid fossa. This would likely lead to arthritic changes at the radioscaphoid joint. Trumble et al. (1986), in a study very similar to the investigation described in this chapter, performed simulated STT fusions with the wrist in 70 to 80 percent of maximal ulnar deviation. This would be expected to bring the scaphoid into greater axial length and may account for their finding of decreased lunate loading following STT arthrodesis. They also used uniaxial electronic strain gauges which were not necessarily oriented in the direction of the principal stresses and thus could not determine the maximum principal or shear strains in the lunate. The use of a rosette strain gauge as done here completely defines the strain field and allows computation of both compressive and shear strains.

Limitations

Potential criticisms of this study include the use of preserved rather than fresh cadaveric limbs and load testing of the wrist in only one position (neutral). Preserved limbs are inherently more stiff and return to the original strain reading following removal of the load was incomplete. However, Trumble et al. (1986) found no significant differences in the strain patterns of fresh versus preserved specimens. They also demonstrated that procedures that decreased lunate load in the neutral position generally continued to be effective in other positions. These specimens also had normal lunates without collapse. In this situation STT arthrodesis with a 45° radioscaphoid angle increases lunate strain. We cannot rule out the possibility that this same fusion might relieve lunate loading in wrists with collapse. Although we did not specifically test radial shortening in this biomechanical study, we feel that biomechanically this should correspond to the same results as obtained from ulnar lengthening.

Conclusions

- Ulnar lengthening of 3 mm (or enough to create neutral variance) decreases both compressive and shear strains on the lunate. Ulnar lengthening of 2 mm gives inconsistent results. Ulnar lengthening of 4 mm tends to increase both compressive and shear strains.

- STT fusion with a radioscaphoid angle of 45° significantly increases both lunate compressive and shear strains.

- CH fusion tends to decrease compressive strain and increase shear strain, but neither attained statistical significance.

- When performing a joint leveling procedure, the amount of radial shortening or ulnar lengthening should be that needed to establish neutral variance. Leaving a negative variance will not unload the lunate and creating a positive ulnar variance increases lunate load and risks ulnocarpal impingement and distal radioulnar joint incongruity.

References

Agerholm JC, Goodfellow JW: Avascular necrosis of the lunate bone treated by excision and prosthetic replacement. J Bone Joint Surg 1963;45B:110-116.

Almquist EE, Burns JR: Radial shortening for the treatment of Kienböck's disease - a 5-10 year follow-up. J Hand Surg 1982;7:348-52.

Armistead RB, Linscheid RL, Dobyns JH, Beckenbaugh RD: Ulnar lengthening in the treatment of Kienböck's.

Bax JC, Sproul JT, Klug MS, Hodges RE: Triskaphe arthrodesis, solution or problem. Presented at the 43rd Annual Meeting of the American Society for Surgery of the Hand. Baltimore, MD, 1988.

Beckenbaugh RD, Shives TC, Dobyns JH, Linscheid RL: Kienböck's disease: the natural history of Kienböck's disease and consideration of lunate fractures. Clin Orthop 1980;149:98-106.

Cave EF: Kienböck's disease of the lunate. J Bone Joint Surg 1939;21:858-66.

Chan KP, Huang P: Anatomic variations in radial and ulnar lengths in the wrists of Chinese. Clin Orthop 1971;80:17-20.

Chuinard RG: Capitate-hamate fusion in the treatment of Kienböck's disease. From Tubiana, The Hand, Philadelphia. WB Saunders Co, 1985, pp1121-1127.

Codega G, Codega O, Kus H: Neurolysis of the median nerve in the carpal tunnel as a surgical treatment of Kienböck's disease. Int Surg 1973;58:378-82.

Danis A: Osteomalacie du semi-luniare traitee par exerese et prothese acrylique. Resultat apres trois ans. Acta Chir Belgica 1951;50:120-126.

Dornan A: The results of treatment in Kienböck's disease. J Bone Joint Surg 1949;31B:518-20.

Eiken O, Niechajev I: Radius shortening in malacia of the lunate. Scand J Plast Reconstr Surg 1980;14:191-6.

Evans G, Burke FD, Barton NJ: A comparison of conservative treatment and silicone replacement arthroplasty in Kienböck's disease. J Hand Surg 1986;11(B)98-102.

Garcia-Elias: Pressure distribution of the wrist after limited intercarpal fusions. Presented at the Wrist Investigator's Workshop, Tampa, FL, 1988.

Gelberman RH, Salamon PB, Jurist JM, Posch JL: Ulnar variance in Kienböck's disease. J Bone Joint Surg 1975;57A:674-676.

Gillespie HS: Excision of the lunate bone in Kienböck's disease. J Bone Joint Surg 1961;43B:245-49.

Graner O, Lopes EI, Carvalho BC, Atlas S: Arthrodesis of the carpal bones in the treatment of Kienböck's disease, painful ununited fractures of the navicular and lunate bones with avascular necrosis, old fracture dislocations of carpal bones. J Bone Joint Surg 1966;48(A):767-74.

Hori Y, Tamai S, Okuda M, Sakamoto H, Takita T, Masuhara K: Blood vessel transplantation to bone. J Hand Surg 1979;4:23-33.

Hulten O: Uber anatomische variationen der handgelenkknochen. Ein beitrag zur Kenntnis der genese zwei ver schiedener mondbeinveranderungen. Acta Radiol 1928;9:155-68.

Kato H, Usui M, Minami A: Long-term results of Kienböck's disease treated by excisional arthroplasty with a silicone implant or coiled palmaris longus tendon. J Hand Surg 1986;11A:645-53.

Kienböck R: Uber traumatische malazie des mondbeins und ihre folgezustande: Entartungsformen und kompressionsfrakturen. Fortschr. Geb Rontgenstr 1910;16:78.

Kraemer BA, Young VL, Weeks PM: Functional results after scaphoid-trapezium-trapezoid fusion. Presented at the 43rd Annual Meeting of the American Society for Surgery of the Hand. Baltimore, MD, 1988.

Lichtman DM, Mack GR, MacDonald RI, Gunter SF, Wilson JN: Kienböck's disease: the role of silicone replacement arthroplasty. J Bone Joint Surg 1977;59(A):899-908.

Lippman EM, McDermott LJ: Vitallium replacement of the lunate in Kienböck's disease. Military Surgeon 1949;105:428-4.

Lunn PG, Ashwell J: The detailed blood supply of pronator quadratus and its relevance to the use of a muscle pedicle bone graft in the carpus. Presented at the combined meeting of the Eastern Mediterranean Hand Society and the British Society for Surgery of the Hand, 1985.

McMurtry RY, Youm Y, Flatt AE, Gillespie TE: Kinematics of the wrist. II. Clinical applications. J Bone Joint Surg 1978;60(A)955-61.

Michon J: Kienböck's disease and complications of injuries to the lunate bone. From Tubiana, The Hand. Philadelphia. WB Saunders Co., 1985;pp1106-1116.

Moberg E: Treatment of Kienböck's disease by surgical correction of the length of the radius or ulna. From Tubiana, The Hand. Philadelphia. WB Saunders Co., 1985;1117-1120.

Nahigian SH, Li CS, Richey DG, Shaw DT: The dorsal flap arthroplasty in the treatment of Kienböck's disease. J Bone Joint Surg 1970;52(A)245-52.

Palmer AK, Werner FW: A biomechanical evaluation of operative procedures performed for the treatment of Kienböck's disease. Presented at the 43rd Annual Meeting of the American Society for Surgery of the Hand, Baltimore, MD, 1988.

Palmer AK, Glisson RR, Werner FW: Ulnar variance determination. J Hand Surg 1982;7:376-379.

Persson M: Pathogenese and behandlung der Kienböckschen lunatummalazia: Der frakturtheorie im lichte der erfolge operativer radiusverkurzung (Hulten) und einer neuen operationsmethode-Ulnaverlangerung. Acta Chir Scand 1945 (Suppl 98).

Ramakrishna B, D'Netto DC, Sethu AU: Long-term results of silicone rubber implants for Kienböck's disease. J Bone Joint Surg 1982;64B:361-363.

Roca J, Beltran JE, Fairen MF, Alvarez A: Treatment of Kienböck's disease using a silicone rubber implant. J Bone Joint Surgery, 1976;58A:373-376.

Rossak K: Druckverhaltnisse am Handgelenk uuter besonderer Berucksichtigung von Frakturmechanismen. Z Orthop 1967;103(Suppl):296-299.

Schuh R, Haasters J, Koob E: GroBenbestimmungen von handwurzelknochen zur Herstellung von alloplastichen prosthesis des mondbienes. Zeitschrift fur Orthopaedie und Ihre Grenzgebiete, 1975;113:479-481.

Stahl, Folke: On lunatomalacia (Kienböck's disease). A clinical and roentgenological study, especially on its pathogenesis and the late results of immobilization.

Stark HH, Zemel NP, Ashworth CR: Use of a hand carved silicone-rubber spacer for advanced Kienböck's disease. J Bone Joint Surgery 1981;63A:1359-1370.

Sundberg SB, Linscheid RL: Kienböck's disease: results of treatment with ulnar lengthening. Clin Orthop 1984;187:43-51.

Swanson AB: Silicone rubber implants for the replacement of the carpal scaphoid and lunate bones. Orthop Clin N Amer 1970;1:299-309.

Tajima T: An investigation of the treatment of Kienböck's disease. J Bone Joint Surg 1966;48A:1649.

Tillery B: Kienböck's disease treated with osteotomy to lengthen ulna. Acta Orthop Scandinav 1968;39:359-68.

Trumble T, Glisson RR, Seaber AV, Urbaniak JR: A biomechanical comparison of the methods for treating Kienböck's disease. J Hand Surg 1986;11A:88-93.

Watson HK, Rogers W: Radial styloid impingement following triscaphe arthrodesis. Presented at the 43rd Annual Meeting of the American Society for Surgery of the Hand. Baltimore, MD, 1988.

Watson HK, Ryu J, DiBella A: An approach to Kienböck's disease: triscaphe arthrodesis. J Hand Surg 1985;10A:179-87.

Youm Y, Flatt AE: Kinematics of the wrist. Clin Orthop 1980;149:21-32.

Chapter 8

Material Properties of Ligaments

M.D. Nowak

Introduction

Although of primary interest to the analysis of wrist stability, prior to the late 1970's, wrist ligaments were only described in terms of relative strength or weakness as noted clinically after trauma (Fahrer 1981; Taleisnik 1985; Taleisnik 1976). Biomechanical investigations generally examined ligament behavior indirectly as an integral part of general wrist biomechanics, such as noting the angle of wrist deflection due to specific loading or placing pressure sensitive film between carpal bones to measure contact pressure (Minami, et al., 1985; Ruby, et al., 1987; Viegas, et al., 1987a. and 1987b.).

In the late 1970's, Mayfield et al. were the first to study the biomechanical properties of isolated wrist ligaments (Mayfield, et al., 1979; Mayfield 1984; Williams, et al., 1979). This study examined the ultimate strength parameters of selected ligaments from both fresh and fresh frozen cadaver subjects.

Recent studies of wrist ligaments have been undertaken by Garcia-Elias et al. (1989a. and 1989b.), and Logan and Nowak (1986; 1987) (Nowak and Logan, 1987; 1988a.; 1988b.; 1989; 1991; Nowak 1988; Mayfield, et al., 1979; Mayfield 1984; Minami, et al., 1985). This chapter will summarize the results of these investigations.

Methods

All three study groups utilized fresh or fresh frozen cadaver wrists. The Nowak data was derived from donors in the 40-64 year old range, with approximately equal numbers of males and females (Nowak 1988; Nowak and Logan, 1991). Mayfield utilized subjects with an average age of 60 years (Williams, et al., 1979), while the mean age of the Garcia-Elias specimens was unknown. The latter two study groups made no distinction as to the number of male and female donors. No embalmed specimens were utilized. Bone-ligament-bone assemblies were dissected for each tested ligament. During dissection, any ligament with visible damage (fiber failure, excessive laxity, calcification) was excluded from the test pool. During testing, some ligament test data were discarded due to technical problems, such as slippage in the test system.

In the Nowak study, multiple bone-ligament-bone assemblies were dissected from each wrist to make the most efficient use of the limited numbers of available cadaver wrists. If a specimen could not be tested immediately after dissection, it was wrapped in a saline soaked towel, sealed in an air-tight bag and placed in a refrigerator. The order of ligament testing was varied to remove order bias, and two test ligaments were analyzed in the fully recoverable range, stored and retested. There were no detectable changes in biomechanical properties (Nowak and Logan, 1991).

The studies of Mayfield et al. and Garcia-Elias et al. utilized bones cemented into cups which were attached to axial testing devices (MTS 810 servohydraulic material testing systems in both studies) (Garcia-Elias 1989a.; Mayfield, et al., 1979; Williams, et al., 1979). The data presented by Nowak utilized a wire-fixation technique (Nowak and Logan, 1991). Two perpendicular holes were drilled in each bone, at right angles to the principal ligament fiber orientation. Stiff wires ($<1\%$ strain at 1000 N applied load) 1.12 mm in diameter were inserted and attached to a Monsanto Tensometer 10 electromechanical axial testing device. The two wires prevented bone rotation during testing. Elongation was determined by crosshead position. A concern with this method is bone viscoelasticity effecting the results. Tests of a scapholunate (SL) specimen with an extensometer attached to the ligament insertion sites demonstrated no difference from crosshead measurements (Nowak 1988).

Prior to testing, initial lengths and cross-sectional areas were determined. Garcia-Elias utilized an electronic digital caliper (Garcia-Elias, et al., 1989a.), while the studies of Mayfield and Nowak employed area micrometers (Mayfield, et al., 1979; Nowak and Logan, 1991), which read in square millimeters. Nowak noted that the horse-shoe shaped cross-section of the SL and lunotriquetral (LT) did not lend themselves to the latter mode of calculation, and vernier caliper measurements were utilized. Each ligament was modeled as two right triangles of material joined at their apex. This method was validated by measuring split SL ligaments with the area micrometer (Nowak 1988; Nowak and Logan, 1991). Initial lengths were measured with vernier calipers.

All ligaments were examined for ultimate strength parameters. Mayfield's studies used an elongation rate of 1 cm/sec for most ligaments examined, which converted to between 50 and 100 percent of ligament length per second (Mayfield, et al., 1979; Williams, et al., 1979). SL ligaments were evaluated at 1 mm/sec, due to its short length (Mayfield, et al., 1979). To eliminate the effect of different strain rates due to varying ligament lengths, Garcia-Elias tested each different ligament at 50% ligament length per second.

Nowak examined ten ligaments at 100 mm/min elongation rate, and four ligaments at a range of elongation rates from 0.5 to 500 mm/min to examine the effect of strain rate (Nowak and Logan, 1987; 1991). These studies also investigated stress-relaxation and permanent deformation by sequential strain testing as opposed to single elongation to failure. Prior to testing, ligaments were preconditioned by cycling in the fully recoverable strain range at 0.5 mm/min elongation rate. Sequential strain analysis consisted of elongation to a set strain level, followed by a period of stress relaxation of at least 15 minutes, a low elongation rate (0.5 mm/min) return to zero applied force and a recovery period of 3-5 minutes. Residual laxity after this recovery period was taken as permanent deformation (not plasticity, since the authors could not determine whether deformation was due to plastic flow to microscopic failure which could not be sensed by the measuring system). Strain testing with residual laxity was followed by a return to initial ligament length prior to continued testing. Permanent deformation was calculated until the first evidence of fiber failure. To investigate the effect of sequential strain testing on ultimate strength, further pools of ligaments were examined in single elongation testing to failure.

The ligaments tested were as follows for each study. The Mayfield work examined the radiocapitate (RC), volar radiotriquetral (VRT), radioscaphoid (RS) and the scapholunate (SL). The study of Garcia-Elias included the flexor retinaculum (FR), palmar hamate-capitate (PHC), palmar capitate-trapezoid (PCTd), palmar trapezoid-trapezium (PTT), palmar capitate-trapezium (PCTm), dorsal trapezoid-trapezium (DTT), dorsal capitate-trapezoid (DCT), and dorsal hamate-capitate (DHC). The Nowak studies utilized the radiolunate (RL), ulnolunate (UL), radioscapholunate (RSL), radial collateral (RCol), RC, SL, lunotriquetral (LT), scaphotrapezium (ST), and the two sections of the "V" or deltoid ligament: capitoscaphoid (CS) and capitotriquetral (CT).

All studies were undertaken at room temperature. Moisture was maintained by spraying the specimens with saline in the Mayfield and Garcia-Elias studies, while Nowak utilized a humidity bag with saline soaked sponges helping to maintain moist conditions (Garcia-Elias, et al., 1989a.; Mayfield, et al., 1979, Nowak and Logan, 1991).

The data presented from the Nowak study represents a pool of at least five successful specimens for any given ligament and test mode. The Garcia-Elias data pools have specimen numbers of at least three for each ligament segment tested. The number of specimens tested for each ligament of the Mayfield work is unknown, but was derived from a base of eleven dissected wrists (the number of ligaments discarded for technical reasons is unknown).

Results

Figure 1 displays the initial dimensions of each ligament tested (Garcia-Elias, et al., 1989a.; Nowak and Logan, 1988b.). Figure 1A describes initial ligament length in millimeters, while Figure 1B presents initial cross-sectional areas in millimeters squared (Garcia-Elias, et al., 1989a.; Mayfield, et al., 1979; Nowak and Logan, 1988b.). The Mayfield data was approximated from published bar charts. In cases where both Mayfield and Nowak examined the same ligament, data are presented with the notation "M" or "N" respectively. This figure, as well as subsequent figures, is separated left to right into clinically described functional groups: extrinsic ligaments attached to the forearm, intrinsic ligaments of the proximal row, those connecting the proximal and distal carpal rows, and distal row intrinsic ligaments.

Figure 2 displays the ultimate strength data. Figure 2A displays applied force in Newtons at ultimate strength (Garcia-Elias, et al., 1989a.; Mayfield, et al., 1979; Nowak and Logan, 1988b.; Williams, et al., 1979), while Figure 2B shows percent strain at ultimate strength (Mayfield 1984; Nowak and Logan, 1988b.). The Mayfield and Garcia-Elias data are based upon single elongation to failure testing. The data of Nowak represent sequential strain testing, with four ligaments also examined by single elongation to failure (Nowak and Logan, 1989). The latter ligaments are displayed as N1 (sequential strain data) and N2 (single elongation to failure). The Nowak stresses are presented as true stresses (force divided by current cross-sectional area). The current cross-sectional area was calculated from strain data and initial ligament volume, assuming a constant volume and even "necking down" along the ligament.

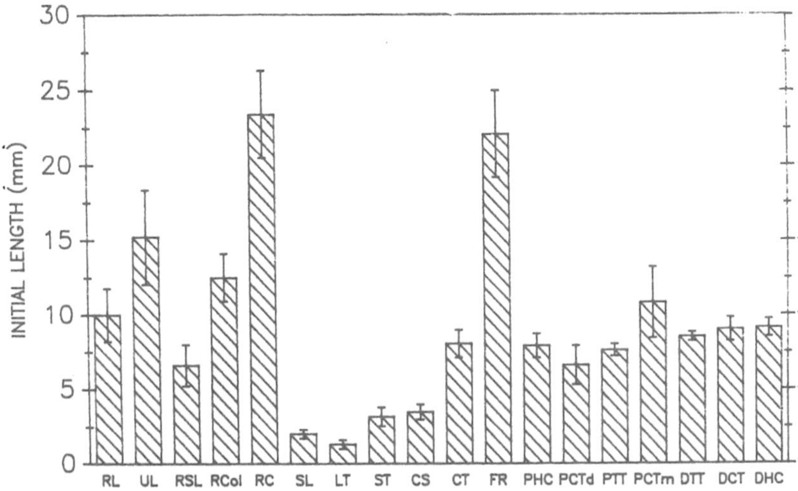

FIGURE 1A. Initial lengths of 18 wrist ligaments, grouped from left to right: extrinsic ligaments, intrinsic ligaments connected to the proximal carpal row, and intrinsic ligaments of the distal carpal row. The distal row data of Garcia-Elias represent dorsal and volar ligament components, while the Nowak data are derived from continuous ligament (some of which include volar and dorsal segments).

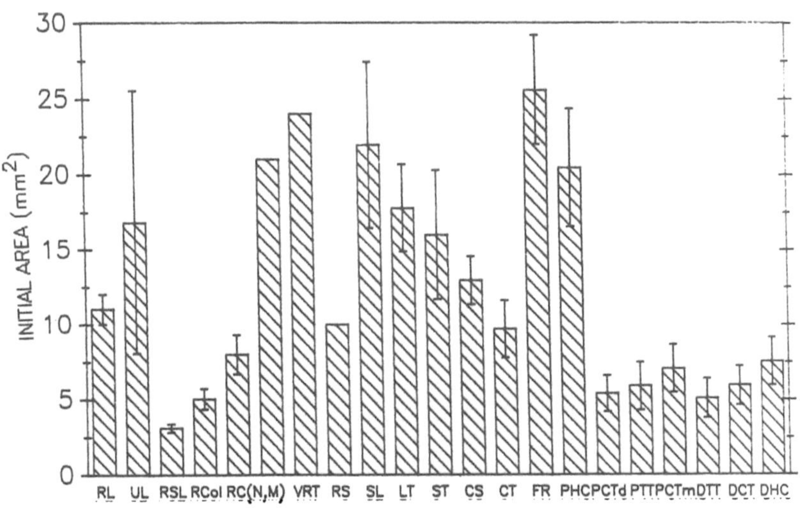

FIGURE 1B. Initial cross-sectional areas of 20 ligaments. In addition to the sources listed in Figure 1, mean data (without standard deviations) from Mayfield's work are listed. In cases where both Nowak and Mayfield examined ligaments, data are presented as (N,M) (Nowak, Mayfield).

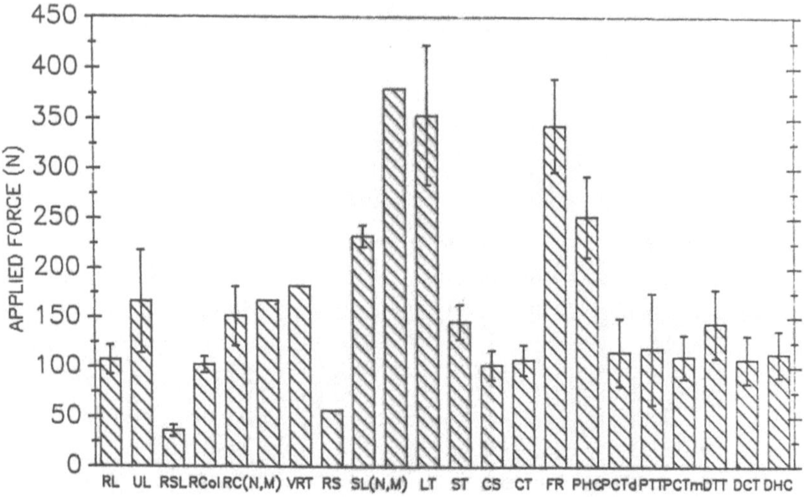

FIGURE 2A. Applied force (N) at ultimate strength. 20 ligaments are listed, with two (RC, SL) being examined by both Nowak and Mayfield. It should be noted that the Mayfield and Garcia-Elias data are from single elongation to failure testing, while the Nowak data were derived from sequential strain level testing.

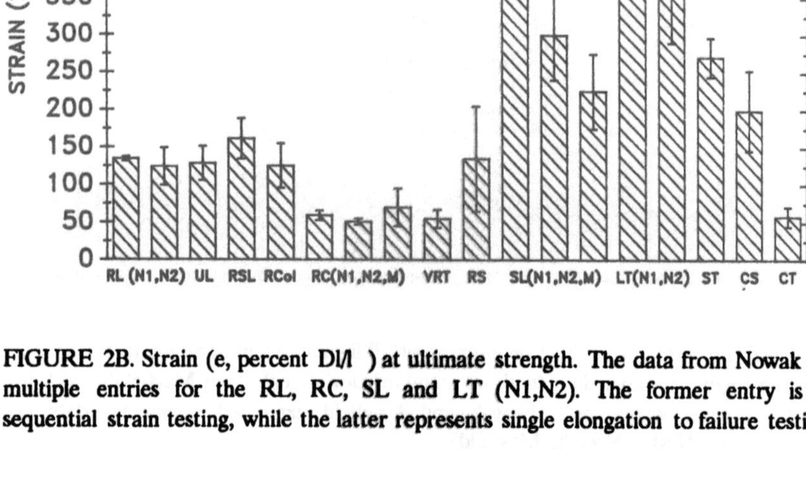

FIGURE 2B. Strain (e, percent Dl/l) at ultimate strength. The data from Nowak list multiple entries for the RL, RC, SL and LT (N1,N2). The former entry is by sequential strain testing, while the latter represents single elongation to failure testing.

Figure 3 displays a typical stress-strain curve for a SL tested by the sequential strain method (Nowak and Logan, 1991). The toe or "J" region, followed by quasi-linearity, yield and ultimate strength is evident, as well as the ability of the multifibered ligament to accept loads beyond ultimate strength.

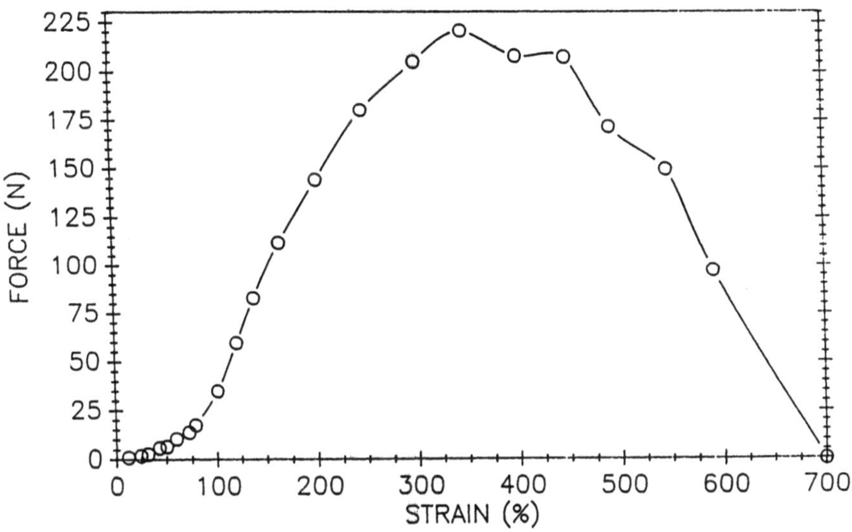

FIGURE 3. Sample of sequential strain testing of a SL at 100 mm/min elongation rate. The toe region, quasilinear region, and ultimate strength points are in evidence. The ligament can accept load at strain levels well beyond ultimate strength.

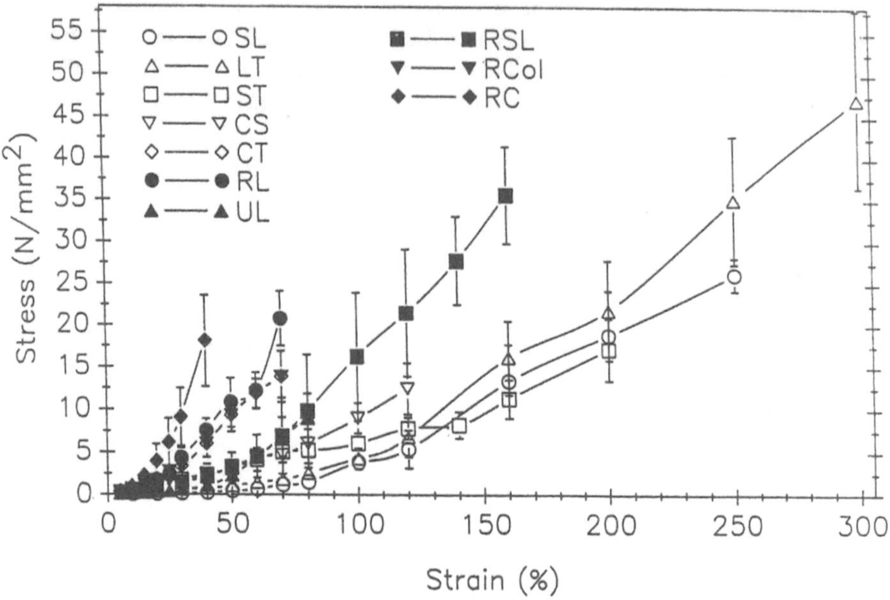

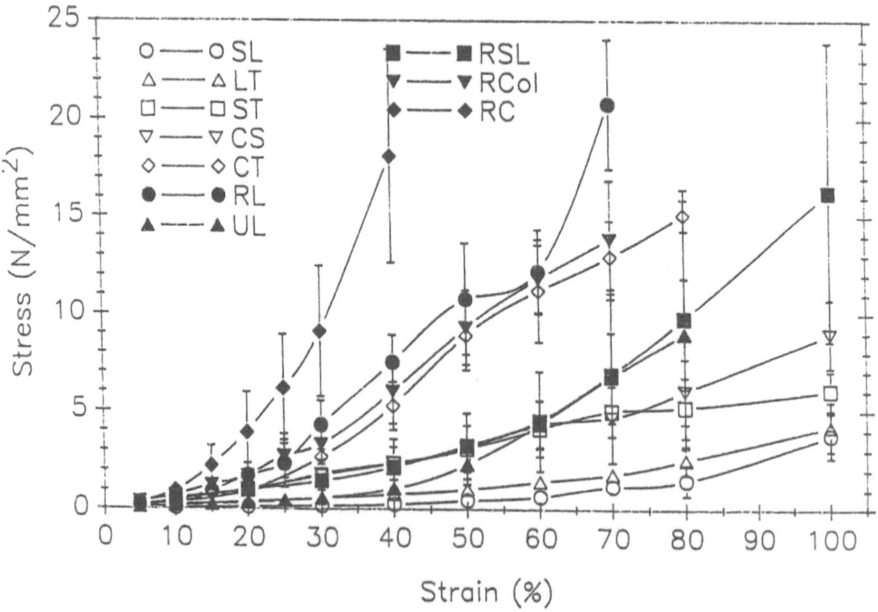

FIGURE 4. Sequential stress-strain data for ten ligaments. Data are presented at sequential strain levels until the first evidence of fiber failure. The intrinsic ligaments are, in general, less stiff than extrinsic ligaments. Figure 4B is the same data, displayed to only 100% strain for ease of comparison.

Figure 4 displays the sequential stress - strain curves for the ten ligaments tested by Nowak (Nowak 1988). Figure 4A displays the curves from zero strain to the first evidence of fiber failure, while Figure 4B displays the same data to a strain level of 100%. For clarity, the intrinsic ligaments are represented by open symbols, and the extrinsic ligaments by closed symbols.

Figure 5 displays the strain at onset of permanent deformation for ten ligaments from the Nowak study at an elongation rate of 100 mm/min (Nowak 1988). This data should be viewed along with Figure 2B, which displays the strain at ultimate strength.

To examine strain rate dependence, Nowak examined four ligaments (SL,LT, RL, and RC) at a variety of elongation rates in the fully recoverable range. Pools of ligaments were examined at either 1 or 100 mm/min elongation rate to failure (Nowak 1988; Nowak and Logan, 1987; 1988a.; 1991). A plot of viscoelastic stress versus sequential strain for the latter testing is shown in Figure 6 (Nowak and Logan, 1991). As can be seen, the extrinsic ligaments are stiffer than the intrinsic ligaments. This effect is not due to the effect of the different strain rates for each ligament, since this would produce nesting of the curves (for example the 1 mm/min SL strain rate falls between the strain rates of 1 and 100 mm/min for the RC, and the stress - strain curves would be expected to follow suit if the differences were due to strain rate alone).

Sequential strain testing at multiple elongation rates allowed for the determination of rate dependent onset of permanent deformation, examination of this region, and evaluation of ultimate strength (Nowak and Logan, 1988a.; 1991). Figure 7 displays the ultimate strength data for all ligaments tested in this manner. Rate dependence can be seen in the applied force at ultimate strength (Figure 7A) as well as percent strain at that level (Figure 7B). Figure 8A displays the permanent deformation as a function of applied strain at two elongation rates for the SL, from onset of permanent deformation to the first evidence of fiber failure. The elongation rate of 100 mm/min presents a maximum deformation of approximately 130% beyond initial, which may place a wrist at risk of pain. The length of an average SL would have increased from 2 mm to approximately 4.6 mm. It should be pointed out that at this strain level there was yet to be any evidence of fiber failure, so the ligament would seem intact. The effect of strain rate is also in evidence as the graph compares data from 100 and 1 mm/min testing. Figure 8B displays the permanent deformation of the RC, which presents a far lower permanent deformation of under 10 percent at 100 mm/min elongation rate. This would convert to a laxity of 2.3 mm in addition to the average initial length of 23.4 mm.

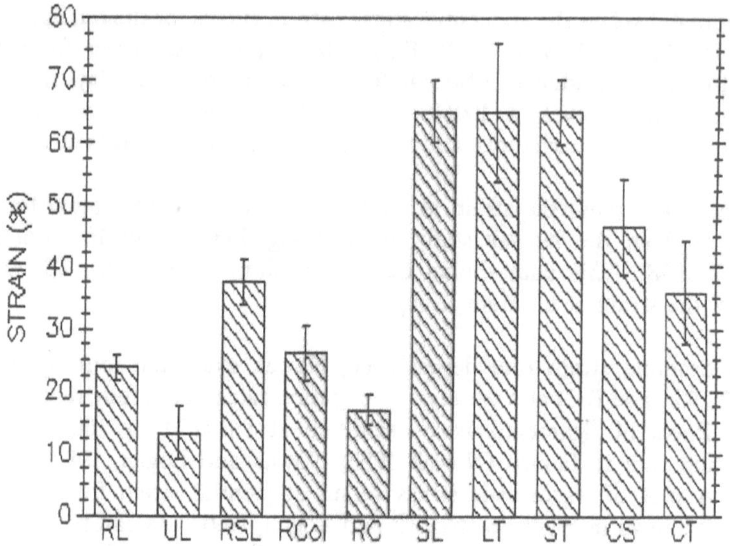

FIGURE 5. Strain at onset of permanent deformation for 10 ligaments. Data derived from sequential strain level testing, and represent the levels above which residual laxity was noted when each ligament was returned to zero applied force.

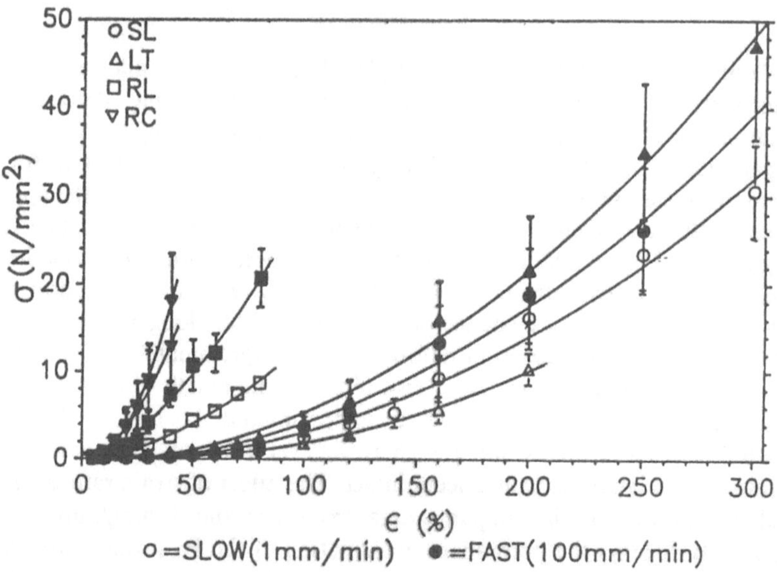

FIGURE 6. Sequential stress-strain data at two elongation rates (1, 100 mm/min) for four ligaments (RL, RC, SL, LT). Data presented at strain levels until the first evidence of fiber failure. Note that strain rates for extrinsic and intrinsic ligaments bracket each other (for example: the slow strain rate for the RC falls between the two values for the SL).

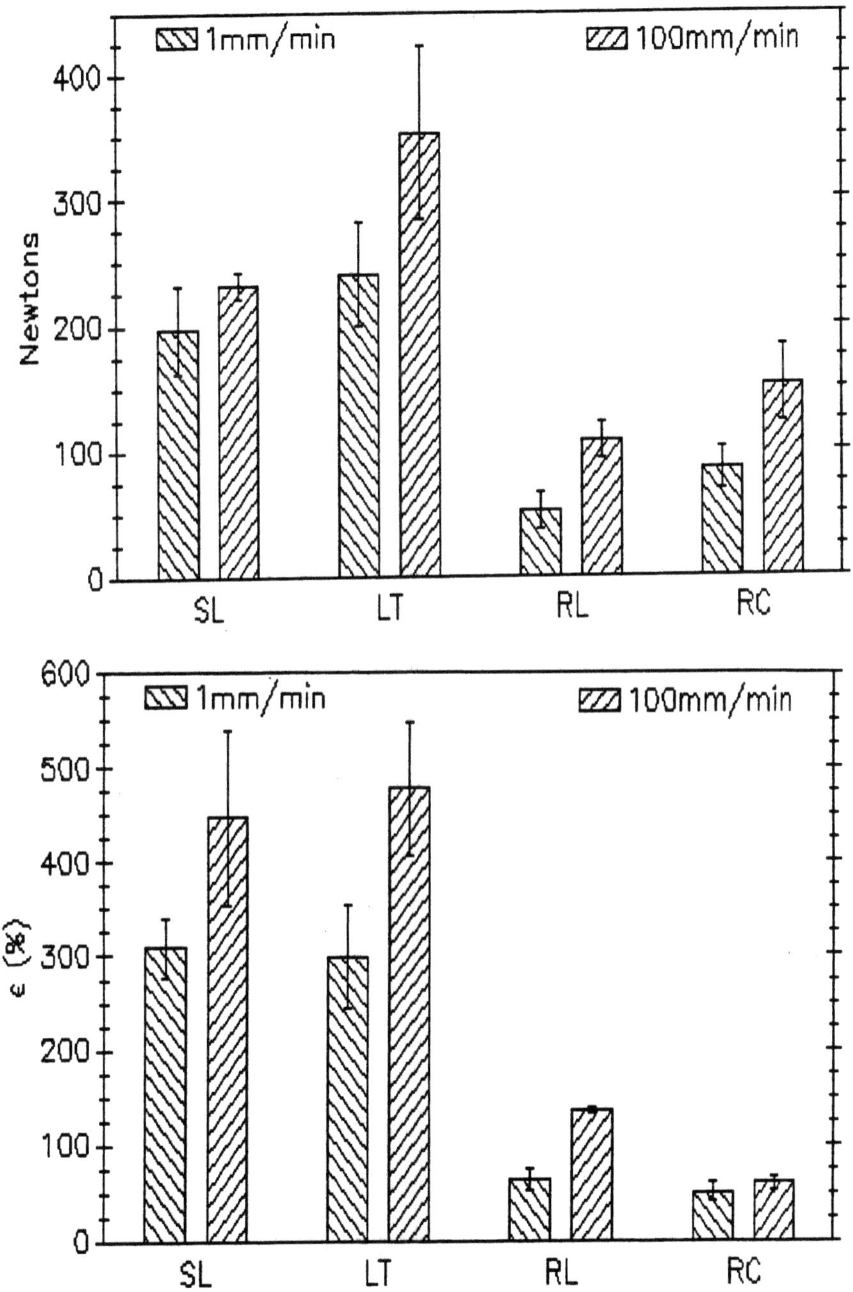

FIGURE 7. Rate dependent ultimate strength data. Figure 7A presents the applied force for 1 and 100 mm/min elongation rates. Figure 7B presents rate dependent strain at ultimate strength.

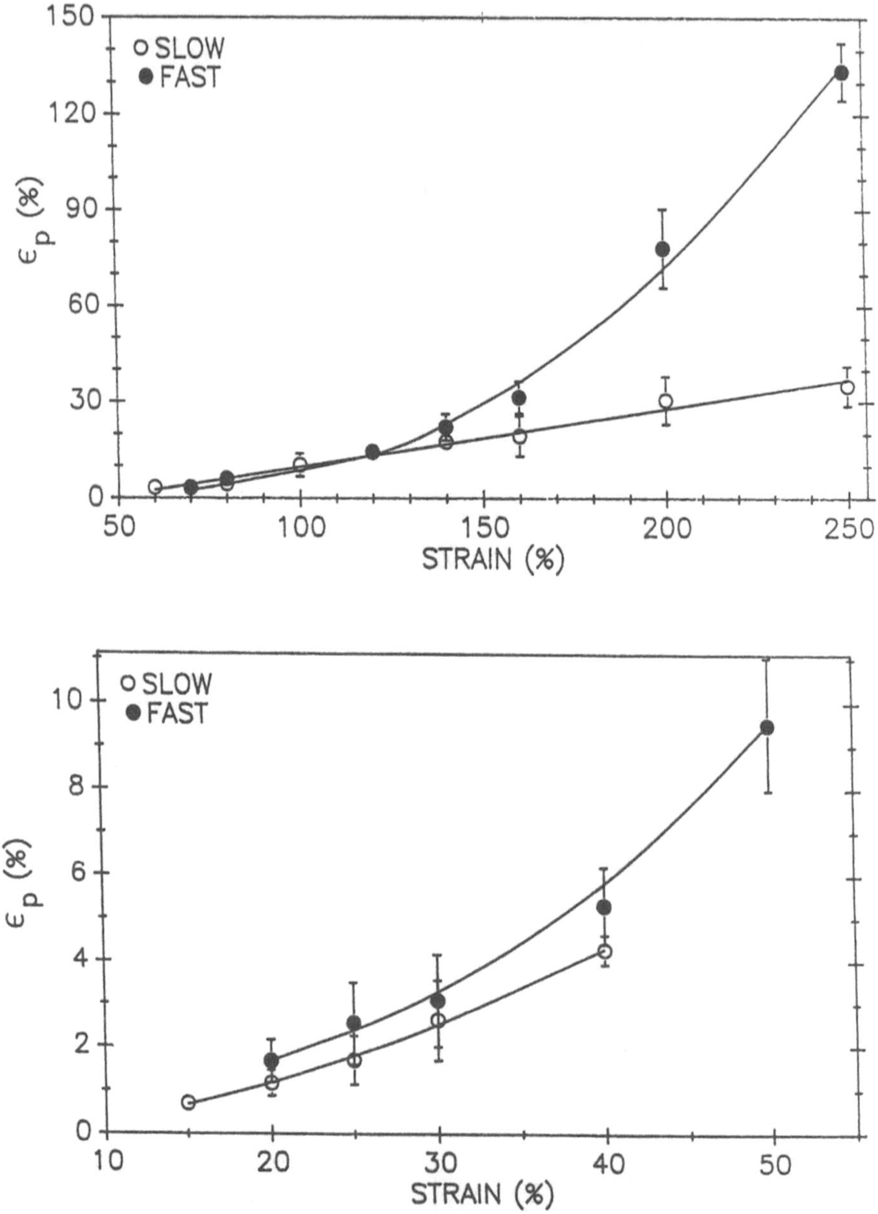

FIGURE 8. Elongation rate dependent permanent deformation. Data points describe the cumulative residual laxity (vertical axis) due to sequential strain testing to a given applied strain level (horizontal axis). Data are presented for elongation rates of 1 and 100 mm/min, from the onset of permanent deformation to the first evidence of fiber failure. Figure 8A (SL) documents the greatest laxity of the ligaments tested, while Figure 8B (RC) presents one of the least deformable ligaments.

Discussion

The study of human wrist ligament biomechanics is valuable both in the basic understanding of wrist function and pathology, along as a required first step in modeling of the wrist. Therefore, the above work can be discussed both on its own and in relation to clinical findings.

To evaluate the relative properties of functional groups of ligaments, along with strain-rate dependence, Nowak compared ten ligaments utilizing one-way ANOVA testing for analysis of variance for more than two sets of data. Ligaments were separated into the clinically used grouping of intrinsic (SL, LT, ST, CS, CT) and extrinsic (RL, UL, RSL, RCol, RC).

Prior to biomechanical evaluation, significant differences were noted for both initial length and cross-sectional area (Figure 1). The intrinsic ligaments were found to be shorter and thicker than extrinsic ligaments ($p<.0001$ in both cases) (Nowak and Logan, 1991).

Onset of permanent deformation did not demonstrate any patterns for applied force or stress, but strain data defined the intrinsic ligaments as being able to accept significantly greater strains prior to measurable permanent deformation (Figure 5) ($p<.0001$) (Nowak and Logan, 1991).

At ultimate strength, intrinsic ligaments as a group accepted greater applied force and true stress ($p<.0230$) as well as significantly higher strain levels (Figure 2) ($p<.0385$). The SL and LT accepted greater force, stress and strain levels than the other sequential strain tested ligaments ($p<.0001$). This last finding may have relevance considering the clinical importance of these ligaments (Nowak and Logan, 1991).

The examination of elongation rate dependance demonstrated significant differences in applied force, stress, and strain at both onset of permanent deformation and ultimate strength (Figures 7 and 8). Using the Student's t-test, the higher elongation rate produced significantly higher values in each category ($p<.0001$) (Nowak and Logan, 1991).

The extrinsic ligaments are often described as less strong than the intrinsic ligaments of the proximal row, and this generally seems to be the case (Figure 2). The intrinsic ligaments of the distal row appear to be strongly connect the carpal bones, and the work of Garcia-Elias bears this out. It should be noted that each pair of distal row carpal bones is connected by at least two of the listed ligament sections. Therefore, total ligament strength (taken as both dorsal and volar components) at ultimate strength is the sum of the parts.

The SL and LT are arguably the most critical ligaments of the wrist, and they accept the greatest load and strain prior to failure (Figures 2 and 7). The reason behind this extended allowable strain is unknown at this time, but clinically may be important since these ligaments may not fully heal (a lax ligament being better than a failed ligament). The ultimate strength behavior as well as the stress strain curves may be of value in developing repair methods which replace the SL and LT with materials which match their unusual (for ligaments and tendons) properties.

Unlike the longer ligaments and tendons, there is evidence of significant strain rate dependence in examination of wrist ligament mechanics (Figures 6 and 7). Clinically, this knowledge may be of interest in the examination of trauma impact, where the tested ligaments can accept greater load and deflection prior to failure.

The permanent deformation regions of the ten ligaments described by Nowak are of interest since the ligaments are intact in this region. A hole or tear would not be in evidence under examination. The long term function of ligaments with permanent deformation is currently being examined.

The effect of permanent deformation on ultimate strength can be seen when comparing data from sequential strain and single elongation testing to failure (Figure 8). The sequential strain data demonstrates higher strains at ultimate strength. However, if permanent deformation is subtracted from the data at each strain level, the corrected strains at ultimate strength are similar to the two test groups (Nowak and Logan, 1989).

The modulus and modeling values displayed in Table 1 present a variety of methods for describing ligament behavior in the quasi-linear stress - strain (as seen as a portion of Figure 3)(Garcia-Elias, et al., 1989a.; Mayfield, et al., 1979; Williams, et al., 1979). Garcia-Elias listed a conversion of elongation to force, while the other data was strain to stress. The latter assumed that cross-sectional areas could be measured or calculated at any given strain level. In examining the data of Nowak, it must be recalled that this is for sequential strain testing and not single elongation to failure. Just as ultimate strength differs between the two cases (Figure 7), the modulus data would be expected to vary.

Table 2 does not list modulus, but rather a power law fit to the sequential strain data (Nowak 1988). This model includes the toe region of the stress - strain curve as seen on the left of Figure 3, as well as the quasi-linear sections. When modeling the ligaments of the wrist, this model may allow determination of behavior in both the physiological (low strain), and low trauma levels of strain (below evident fiber failure).

TABLE 1. Mean modulus and stiffness data for wrist ligaments. Mayfield and Nowak data converts strain (%) to strain (N/mm), while the Garcia-Elias values convert elongation (mm) to applied force (N). The Nowak data is derived from mean pooled data for sequential strain testing of each ligament. All data describes the slope of the quasilinear region of the total ligament behavior.

	MAYFIELD				*NOWAK*									
Ligament	*RC*	*VRT*	*RS*	*SL*	*RL*	*UL*	*RSL*	*RCol*	*RC*	*SL*	*LT*	*ST*	*CS*	*CT*
Modulus (N/mm)	75	74	22	45	27	20	31	26	72	15	22	7	12	25

	GARCIA-ELIAS							
Ligament	*FR*	*PHC*	*PCTd*	*PTT*	*PCTm*	*DTT*	*DCT*	*DHC*
Stiffness (N/mm)	132	216	74	66	78	114	108	86

TABLE 2. Nowak modeled the sequential stress - strain data for zero strain to the first evidence of fiber failure by a power law fit. The data presented is for the equation s= Ae , with A being in N/mm. The correlation coefficient for all ligaments was greater than 0.99.

Ligament	*RL*	*UL*	*RSL*	*RCol*	*RC*	*SL*	*LT*	*ST*	*CS*	*CT*
A	39.1	17.0	15.2	25.3	145.0	3.8	5.5	6.5	9.1	20.4
B	1.9	2.8	1.8	1.6	2.3	2.4	2.0	1.3	1.5	1.4

The Nowak data utilized true stress, based upon a calculated current cross-section (as stated earlier). While not an ideal situation, this method is reasonable if it is recalled that even short ligaments are made of a large number of thin long fibers. These fibers are then assumed to be of constant volume during testing and narrow evenly during extension. An alternate method is to measure engineering stress, which is force divided by initial cross-sectional area. While the latter system is reasonable for materials that reach ultimate strength at similar strain levels, it incorporates significant divergence if strain levels vary by large amounts (as occurred for the described ligaments).

Due to limited sample sizes, none of the studies evaluated specific age or sex related variances in mechanical properties. Nowak noted that visually, individual subject ligaments seemed to be scattered throughout the data pool.

Future Studies

The data presented in this chapter suggest a number of areas for future examination of wrist ligament mechanics. Further testing, to include many of the wrist ligaments not examined to date, is thus called for.

This chapter grouped ligaments into extrinsic and intrinsic functional groups as is done clinically. Based upon the current work and future biomechanical and kinematic studies, a more appropriate grouping of like-behavior ligaments may by useful to the understanding of wrist function.

The ligaments described were all examined axial to their primary fiber orientation, although a functional wrist exerts loadings along a variety of axes. Once a complete three-dimensional understanding of relative carpal motion is developed, work can progress towards examination of ligaments elongated as they would be clinically, while retaining the accuracy and reproducibility that bench testing of individual ligaments affords.

Rate dependence and permanent deformation are areas for continued research. If wrist ligament behavior differs from that noted for longer ligaments (such as those of the knee), then this must be quantified in order to evaluate possible replacements for damaged ligaments.

Similarly, permanent deformation analysis may be of value in determining how best to repair a ligament which may be lax but otherwise intact. If this remains a viable ligament, a repair may be possible that does not require the removal of a functional ligament by fusion or replacement. The sequential strain testing described does not examine the situation of multiple sub-ultimate strength traumas, nor single elongations into the mid or upper levels of the permanent deformation regions.

A useful technical addition to wrist ligament biomechanical testing will be a method for measuring current cross-sectional areas during elongation, upon which to base true stress calculations. Current optical methods may not be accurate when examining the SL and LT which have horse-shoe shaped cross-sections.

Two additional questions not evaluated in the above works are those of sex and age dependence. Due to technical considerations, most cadaveric materials are obtained from older donors, making age dependent testing difficult, although of interest.

In concurrence with future biomechanical testing, histological and biochemical analyses of ligaments will be needed to determine why the individual ligaments behave as they do. This work will be of particular interest for the SL and LT, which seem to present behavior divergent to the longer, less critical ligaments.

In closing, the work presented here forms a basis upon which to develop future strategies for the mechanical evaluation of human wrist ligaments. The future holds many questions to be answered, both in the mechanical behavior of wrist ligaments, and the understanding of why this behavior occurs.

References

Fahrer M: Introduction to the anatomy of the wrist. In: Tubiana R ed. The hand. WB Saunders Co., Philadelphia 1981;pp.130-152.

Garcia-Elias M, An KN, Cooney WP, Linscheid RL, Chao EYS: Stability of the transverse carpal arch: an experimental study. J Hand Surg 1989a.;14A:277-282.

Garcia-Elias M, An KN, Cooney WP, Linscheid RL, Chao EYS: Transverse stability of the carpus: an analytical study. J Orthop Res 1989b.;7:738-743.

Logan SE, Nowak MD: Intrinsic and extrinsic wrist ligaments: biomechanical and functional differences. Biomed Sci Instrum 1987;23:9-13.

Logan SE, Nowak MD, Gould PL, Weeks PM: Biomechanical behavior of the scapholunate ligament. Biomed Sci Instrum 1986;22:81-85.

Mayfield JK: Patterns of injury to carpal ligaments. Clin Orthop Rel Res 1984;187:36-42.

Mayfield JK, Williams WJ, Erdman AG, Dahlof WJ, Wallrich MA, Kleinhenz WA, Moody NR: Biomechanical properties of human carpal ligaments. Orthop Trans 1979;3:143-144.

Minami A, An KN, Cooney WP, Linscheid RL, Chao EYS: Ligament stability of the metacarpophalangeal joint: a biomechanical study. J Hand Surg 1985;10A: 255-260.

Nowak MD: Wrist ligament biomechanics. D Sc thesis, Washington University, 1988.

Nowak MD, Logan SE: Distinguishing biomechanical properties of intrinsic and extrinsic human wrist ligaments. J Biomech Engin 1991 (in press).

Nowak MD, Logan SE: Wrist ligament ultimate strength: single elongation versus sequential strain testing. Proc 15th Ann Northeast Bioeng Conf 1989;pp.213-214.

Nowak MD, Logan SE: Strain rate dependent permanent deformation of human wrist ligaments. Biomed Sci Instrum 1988a.;24:61-65.

Nowak MD, Logan SE: Ultimate strength parameters of 10 clinically significant human wrist ligaments. IEEE Eng in Med & Biol Soc 10th Ann Int Conf 1988b.;pp.718-719.

Nowak MD, Logan SE: Rate dependent biomechanics of four clinically significant human wrist ligaments. Adv Bioeng 1987;7-8.

Ruby LK, An KN, Linscheid RC, Cooney WP, Chao EYS: The effect of scapholunate ligament section of scapholunate motion. J Hand Surg 1987;12A:767-771.

Taleisnik J: The wrist. Churchill Livingston, New York 1985;pp.13-38.

Taleisnik J: The ligaments of the wrist. J Hand Surg 1976;1:110-118.

Viegas SF, Tencer AF, Cantrell J, Chang M, Clegg P, Hicks C, O'Meara C, Williamson JB: Load characteristics of the wrist. Part I. The normal wrist. J Hand Surg 1987a.;12A:971-977.

Viegas SF, Tencer AF, Cantrell J, Chang M, Clegg P, Hicks C, O'Meara C, Williamson JB: Load characteristics of the wrist. Part II. The normal wrist. J Hand Surg 1987b.;12A:978-985.

Williams WJ, Mayfield JK, Erdman AG, Dahlof WJ, Wallrich MA, Kleinhenz WA, Moody NR: Biomechanical properties of human carpal ligaments. Trans Ann Meeting Orthop Res Soc 1979;26.

Chapter 9

Muscle Function

K-N. An, E. Horii, and J. Ryu

Introduction

When skeletal muscle is stimulated, it is rapidly activated and changed from a passive tissue into a dynamic tissue capable of developing force. The potential force which the muscle can possibly generated depends on the size and architecture of the muscle. In addition, the potential force generation is a function of the velocity of shortening or lengthening and the muscle length at the time of contraction. Several important biomechanics parameters have been defined to characterize the physiology of muscle. The definition and significance of these parameters and available data associated with the wrist joint will be reviewed in this chapter. Tendon excursion and moment arm of the muscle and tendon around the joint will further determine the effectiveness of the muscle in providing resistance and movement of the adjacent limb segments.

Mechanics of Muscle Physiology

Size

Three parameters have commonly been used to describe the size and morphology of the muscle (Brand, et al., 1981; An, et al., 1981). Muscle fiber length (FL) is related to the potential for tendon excursion. The physiological cross-sectional area (PCSA) of a muscle is proportional to the maximum tension of the muscle. Since work equals force times distance, the muscle mass or volume is proportional to total work capacity.

A systematic and consistent method of measurement has been developed by Brand et al. (1981), although it was anticipated by the drawings of a geometric model of muscle fibers by Nicolaus Steno (1667). The hand and elbow was first positioned with all joints in the neutral or resting position and held in that position throughout, using Kirschner wire fixation when necessary. For each muscle, the origin and insertion of the muscle on the tendon were identified. The tendon of insertion was sectioned distal to the muscle fibers and allowed to rotate freely about the tendon or bone of origin without alteration of the fibre length. During this rotation, the muscle fibers served as radii of an arc of a circle whose center was the point of origin of the fibre and the circumference traced by the fibre attachment to the tendon. With careful manipulation, the muscle formed the shape of a parallelopipedon. The muscle fibers were then seen to be almost uniform and parallel. Their length were then measured directly with the individual fibers lying obliquely between origin and tendon of insertion. To check that no traction had been applied to the muscle fibre during the measurement, the tendon was then rotated back to its original position. If there was now found to be no overlap between the cut ends of the tendon, that overlap was deducted from the measured fibre length. The muscles were then dissected free and their volume measured by a water displacement technique. The physiological cross-sectional area was calculated by dividing volume by mean fibre length.

Data for these three parameters of muscle size have been reported by several investigators (Brand, et al., 1981; An, et al., 1981; Lieber, et al., 1990). The ranges of these reported data are summarized in Table 1-3 for muscle volume, fibre length and PCSA. Five wrist muscles are included: flexor carpi radialis (FCR), flexor carpi ulnaris (FCU), extensor carpi radialis brevis (ECRB), extensor carpi radialis longus (ECRL), and the extensor carpi ulnaris (ECU). Due to some observed variations among these reported data, the ratio of these value with reference to those of FCR are also included.

TABLE 1. Muscle Volume

Muscle	*Volume (cm^3)*	*Ratio (x FCR)*
FCR	10.9 - 12.4	1.0
FCU	15.2 - 15.4	1.23 - 1.41
ECRB	13.8 - 15.8	1.21 - 1.27
ECRL	11.8 - 18.3	1.08 - 1.55
ECU	13.6 - 14.9	0.95 - 1.25

TABLE 2. Muscle Fibre Length

Muscle	*Length (cm)*	*Ratio (x FCR)*
FCR	5.1 - 5.8	1.0
FCU	4.2 - 4.8	0.82 - 0.83
ECRB	4.8 - 6.1	0.91 - 1.20
ECRL	7.6 - 9.3	1.34 - 1.82
ECU	4.5 - 5.1	0.78 - 1.00

TABLE 3. Muscle Physiological Cross-sectional Area

Muscle	*Area (cm^2)*	*Ratio (x FCR)*
FCR	2.0	1.0
FCU	3.2 - 3.4	1.60 - 1.70
ECRB	2.7 - 2.9	1.02 - 1.45
ECRL	1.5 - 2.4	0.75 - 1.20
ECU	2.6 - 3.4	1.10 - 1.70

Architecture

The arrangement of the muscle fibre architecture will further influence the characteristics of the muscle contraction. It has been demonstrated that parallel muscle fibers produce a length-tension curve with maintained force throughout a greater distance than the sharply peaked length-tension curve of shorter fiber pennate muscles (Woittiez, et al., 1984; Kaufman, et al., 1989). The muscle architecture has been defined based on the pennation angle. More recently, the concept of muscle index of architecture, i_a, has been adopted to describe the muscle architecture and the length-tension relationship. As shown in the schematic representation of the muscle (Figure 1), the muscle fibers lie at an angle to the direction of induced motion. Two measurements can be taken at muscle optimum length, the muscle fiber length and the muscle belly length. The ratio of the mean fiber length to the muscle belly length is defined as the index of architecture (Woittiez, et al., 1984). A typical length-tension relationship of muscles as a function of different architecture indices is shown in Figure 2.

TABLE 4. Pennation Angle and Index of Architecture

Muscle	*Pennation Angle (degrees)*	*Index of Architecture*
FCR	3.1	0.31
FCU	12.1	0.19
ECRB	8.9	0.38
ECRL	2.5	0.82
ECU	3.5	0.28

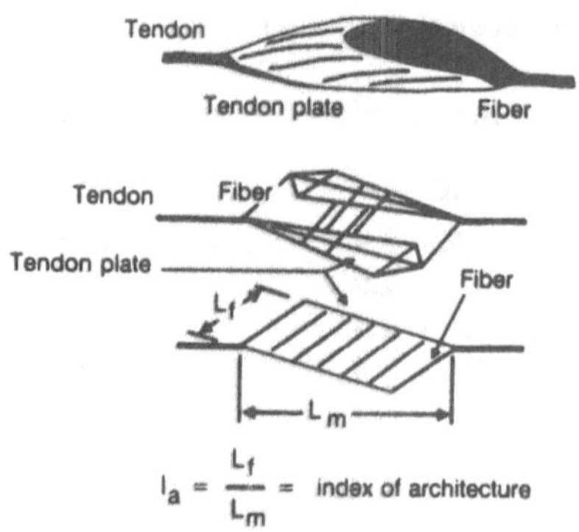

FIGURE 1. Schematic representation of muscle fiber arrangement and architecture. Important architectural parameters of muscle fiber length (L_f) and muscle belly length (L_m) define the index of architecture, i_a.

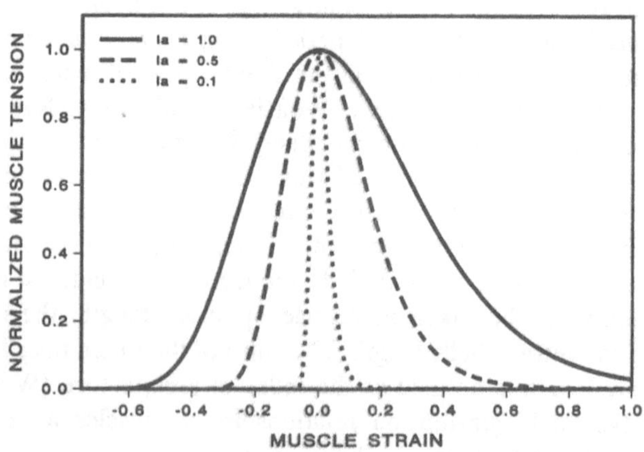

FIGURE 2. Normalized active length-tension relationship of muscle with differing architecture. Index of architecture is indicated.

The pennation angles and the architecture indices for the wrist muscle have been measured by Lieber et al. (1990) by using laser diffraction. The five muscles differ significantly in these parameters (Table 4).

Mechanics of Muscle Function

Prime Wrist Mover

In moving the finger or wrist, each tendon across the wrist joint slides a certain distance to execute the movement. Knowledge of the amplitude of tendon excursion has a significant application to hand surgery. In addition, such data provide insight into the mechanics of muscle function. It has been demonstrated that from the tendon excursion and angular displacement curves, the slope at various joint angles throughout the range of joint motion can be derived. This represents the moment arm about the center of rotation of that particular tendon in the plane of motion (Landsmeer 1961; An, et al., 1983; An, et al., 1984). The tendon excursions of wrist muscles during hand and wrist motion has been examined by numerous investigators as well. (Armstrong, et al., 1977; Brand 1985; Horii, et al., 1990; Ohnishi, et al., 1990).

In one experiment, the tendons of the five prime wrist motors (ECRL, ECRB, ECU, FCR, and FCU) were sectioned 7 cm proximal to the radiocarpal joint. The humerus of the specimen was firmly fixed on a loading frame with the elbow flexed to 90°. The forearm was fixed in an acrylic plastic tube by a steinman pin drilled into the radius, allowing stable positioning of the forearm in various degrees of prono-supination. The proximal end of the sutures securely attached to each tendon were wrapped around the hub of an electric potentiometer for the measurement of tendon excursions. A weight of 200 g was applied to each of the wrist flexor and extensor tendons to eliminate slackness. A plastic bar was inserted into a hole drilled in the intramedullary shaft of the third metacarpal to provide external control of wrist motion. An electromagnetic tracking device (3-space Isotrak, Polhemus Navigation Sciences, P.O.Box 560, Colchester, Vermont, 05446) was used to record three-dimensional wrist motion, defined by the position and orientation of the sensor on the third metacarpal with respect to the distal radius, where source was attached.

The wrist joint was passively moved from full flexion to full extension (FEM) and full radial deviation to full ulnar deviation (RUD) with the forearm in different supination-pronation rotation. During wrist motion, the analog signals from the electric potentiometer, which defined tendon excursion and the signals from the tracking device, which defined wrist motion, were simultaneously recorded by the digital computer. The excursion, as a function of the angular displacement, was then obtained for each tendon. The slope of these curves was analyzed to obtain the effective moment arm of the tendon.

TABLE 5. Tendon excursion of the prime wrist motor tendons. (mm for 100° of flexion extension motion, and 50° of radioulnar deviation) (n=7).

	Flexion Extension	*Radioulnar Deviation*
	mean ± SD	*mean ± SD*
ECRL	12 ± 3	17 ± 1
ECRB	20 ± 3	11 ± 1
ECU	10 ± 2	16 ± 2
FCR	25 ± 4	7 ± 1
FCU	28 ± 4	12 ± 3

TABLE 6. Moment arms of prime wrist motor tendons.(mm) (n=7)

	Flexion Extension	*Radioulnar Deviation*
	+: Extensors *-: Flexors*	*+: Ulnar deviators* *-: Radial deviators*
	mean ± SD	*mean ± SD*
ECRL	+ 7 ± 2	- 19 ± 2
ECRB	+ 12 ± 2	- 13 ± 2
ECU	+ 6 ± 1	+ 17 ± 3
FCR	- 15 ± 3	- 8 ± 2
FCU	- 16 ±	+ 14 ± 3

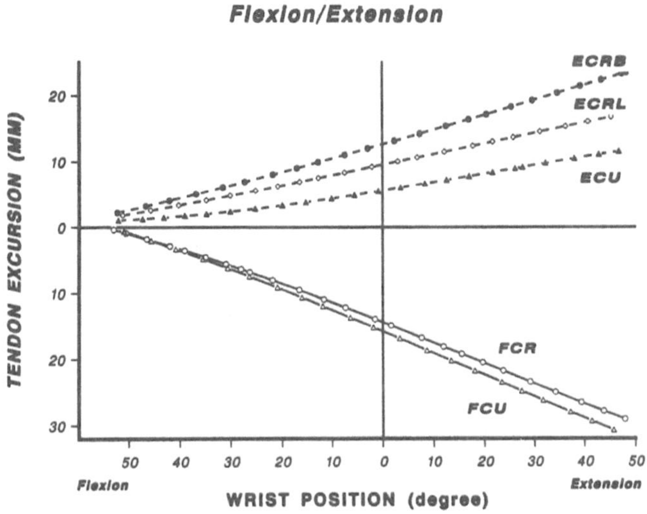

FIGURE 3. Tendon excursion of the five wrist motor tendons during flexion extension motion.

The typical patterns of tendon excursion during wrist flexion/extension and radial/ulnar deviation are shown in Figure 3 and 4 respectively. The averaged amplitude of tendon excursions during 100° of FEM and 50° of RUD in forearm neutral rotation are listed in Table 5. The averaged moment arms of each tendon computed are shown in Table 6. During FEM, the FCR, FCU, and ECRB provided larger tendon excursions, than those of the ECRL and ECU. During RUD, the ECRL and ECU had larger excursions, while that of the FCR was quite small.

Forearm rotation did not produce any significant difference regarding the wrist rotation axes. Significant changes were noted only for excursions of the ECU tendon. In the pronated position, the excursion of ECU decreased during FEM and increased in RUD. The other four tendons did not show any significant changes in moment arm or tendon excursion.

The results clearly demonstrate that the prime muscles are the FCR and FCU for flexion, the ECRB for extension, the ECRL for radial deviation, and the ECU and FCU for ulnar deviation. The FCR does not behave as a significant radial deviator. The presence of the pisiform bone provides a greater flexion moment arm for the FCU tendon compared to the FCR, analogous to the effect that the patella has on the quadriceps tendon. The functional difference between the wrist tendons in each direction of wrist motion should be understood for accurate repair, reconstruction and transfer of the wrist motor tendons.

In spite of three dimensional orientation of the wrist tendons to the rotation axes and the complexity of carpal bone motion, the moment arms for wrist motions were maintained fairly consistently and correspond well with the anatomic location of the tendons. These findings are related to several important anatomical considerations: 1) The extensor retinaculum keeps a consistent relationship of the wrist extensors to the rotation axes, 2) the FCR is firmly fixed in a fibro-osseous groove formed in part by tuberosity of the scaphoid and trapezium, and 3) FCU inserts on the pisiform. It has been postulated, but yet to be proved, that flexion of the scaphoid during wrist FEM or RUD may carry the FCR volarly and thus affect the moment arm of the FCR with these motions (Agee).

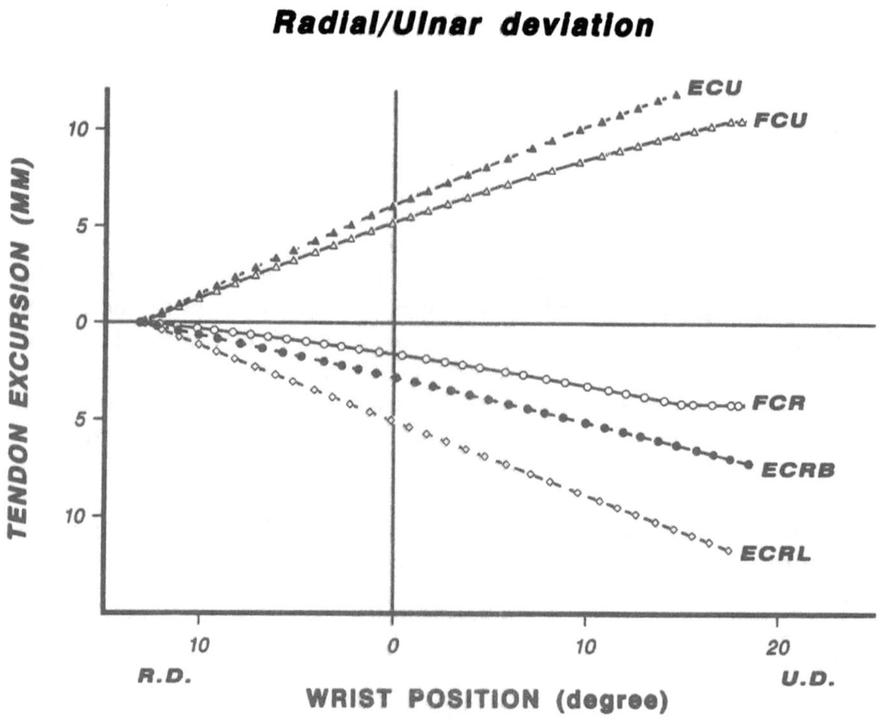

FIGURE 4. Tendon excursion of the five wrist motor tendons during radioulnar deviation.

The effect of various tendon transfers about the wrist can be directly assessed utilizing relatively simple experimental protocols. For example, the transfer of extensor carpi radialis longus (ECRL) to extensor carpi ulnaris (ECU) has been proposed as an effective treatment option to counter radial deviation of the wrist and metacarpals, a condition associated with the development of ulnar drift of the fingers. A recent biomechanical study was completed where cadaver specimens were tested in an apparatus that measured a) the forces acting on the hand to restrain it in seven characteristic wrist configurations, and b) the amount of hand pronation/supination that occurred as a result of loads applied to the tendons of the six major wrist muscles (Berger, et al., 1988). Each specimen was tested with the ECRL tendon intact, surgically released, and transferred to the insertion point of ECU. In the intact and transferred states, the ECRL tendon was loaded sequentially while the remaining five wrist tendons were subjected to equal constant loads. In all three experimental ECRL test states, forces were also applied to all intact wrist tendons in a manner designed to represent physiologic load sharing. When the ECRL tendon was loaded sequentially, the transfer resulted in the predictable increase in the radially directed restraining force and the predictable supination of the hand relative to the forearm. When all intact tendons were loaded physiologically, the transfer also resulted in an increase in the radially directed restraining force. Significant differences between test states occurred generally only between the intact and release states of the ECRL tendon and not between release and transferred states. The results confirm that the ECRL-to-ECU tendon transfer procedure leads to forces and displacements that tend to correct the undesired deformities commonly associated with advanced rheumatoid wrist disease; however, the similarity of results in release and transferred states indicates that the effectiveness of the procedure involves the loss of normal ECRL function rather than reinsertion into the ECU per se.

Finally, in combining the information of the tendon moment arm and the muscle PCSA, the potential moment contribution of each muscle estimated could be established, (Figure 5) providing insight into the normal moment balance at the wrist joint. When a tendon transfer or soft tissue reconstruction is planned, the entire tendon balance must be considered. For example, the FCU tendon is recommended for reconstruction of radial nerve palsy (Green, 1988). However, the moment contribution of this muscle is quite large (Volz, et al., 1980), and it is easily predicted that an imbalance of wrist might result after the FCU transfer. Transfer of the FCR instead of the FCU to maintain the strong stabilizing effect of the FCU for wrist flexion and ulnar deviation, as well as to take advantage of the greater potential tendon excursion, is biomechanically more favorable.

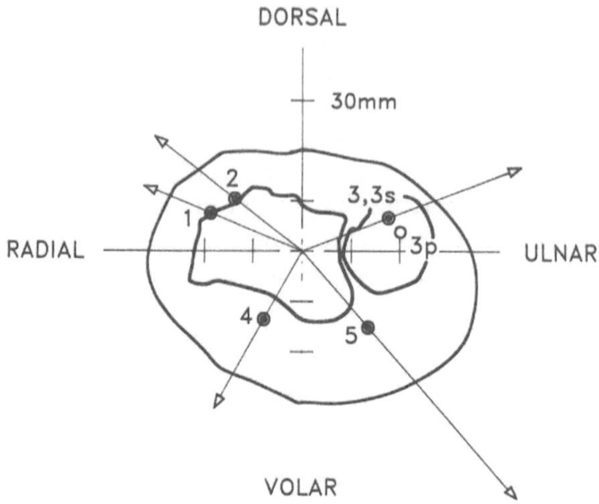

FIGURE 5. The mean values of calculated moment arm were expressed as dots with numbers on the picture of cross-section of the wrist joint. 1: ECRL; 2: ECRB; 3: ECU in neutral wrist rotation; 3p: ECU in pronation position; 3s: ECU in supination position; 4: FCR; 5: FCU. The potential moment contribution of five wrist motor tendons to the wrist joint was estimated by multiplying the moment arm of the muscle by its physiological cross-sectional area, and expressed by arrows.

Finger and Thumb Muscles

Tendons of finger and wrist muscles across the wrist joint would provide mechanical balance of the wrist joint as well. All thumb motors and their corresponding moment arms, with the wrist and forearm in a neutral position, are displayed in Figure 6.

Extensor pollicis longus (EPL) shows relatively weak extensor capabilities (4.0 - 8.3 mm). The values were smaller in supination (4.0 - 7.5 mm) verses neutral (6.0 - 7.8 mm) or pronated positions (6.3 - 8.3 mm). Wrist FEM position had little influence on the moment arm. Irrespective of forearm position, the values increased as the wrist ulnarly deviated. Recorded moment arms were 7.5 - 8.3 mm (ulnarly deviated), 6.1 - 7.8 mm (neutral), and 4.0 - 7.5 mm (radially deviated). The redial moment arm ranged from 15.3 - 17.9 mm. the value was the greatest in supination, and smallest in pronation.

Flexor pollicis longus (FPL) shows a moment arm ranging from 10.7 - 17.0 mm in flexion, largest in supination (14.8 - 17.0 mm), and smallest in pronation (10.7 - 11.0 mm). No large change was observed with changes of the wrist positions. The radial moment arm was small, ranging from 0.9 - 3.0 mm with the lowest occurring in supination.

Abductor pollicis longus (APL) shows a small flexor moment arm ranging from 4.8 - 8.3 mm. Forearm position values were 4.8 - 6.1 mm (pronation), 6.0 - 6.8 mm (neutral), and 4.8 - 8.3 mm (supinated). No changes occurred with wrist motion. APL has large a radial moment arm (21.7 - 23.6 mm). This value was greatest in pronation. Extensor pollicis brevis (EPB) shows itself to be a weak flexor (1.6 - 5.4 mm), weakest in pronation (1.6 - 3.0 mm), and similar in supination (4.6 - 5.3 mm) and neutral positions (3.8 - 5.4 mm). Changing the position of the wrist did not alter the values significantly. Extensor pollicis longus (EPL) is a strong radial deviator (23.6 - 25.3 mm).

Finger mover moment arms are illustrated in Figure 7. The scale of this figure has been expanded to more clearly show the locations of these tightly grouped tendons. Extensor disitorium communis (EDC) 2-4 showed moment arms ranging 12.4 - 18.5 mm in FEM. The values of EDC 3, 4 were greater in supination verses neutral or pronation, and also were greater with the wrist radially deviated. The values were at their lowest with the wrist ulnarly deviated. the moment arms of EDC 2 (13.7 - 15.6 mm) and EIP (13.3 - 17.1 mm) did not change significantly with position of the forearm and the wrist. EDC 2, 3, 4 and extensor iudicis proprius (EIP) have small RUD moment arms. EDC 2 (3.4 - 5.9 mm) and EIP (2.2 - 3.3 mm) act as radial deviators and EDC 3 and 4 (1.4 - 5.4 mm) act as ulnar deviators for all forearm and wrist positions.

Extensor digiti minimus (EDM) has moment arms ranging from 9.2 - 16.5 mm in FEM. The values were greater with forearm supination (11.3 - 16.5 mm), and smaller in pronation (9.2 - 13.0 mm). the values were also greater with radial deviation (13.0 - 16.5 mm). Maximum EDM moment arm (16.5 mm) was obtained with the wrist radially deviated and the forearm supinated, while the minimum value (9.2 mm) occurred with the wrist ulnarly deviated and the forearm pronated. the moment arms in RUD ranged 14.4 - 17.0 mm. The value was largest with pronation.

For FEM, flexor digitorium superficialis (FDS) moment arms were larger than flexor digiforum profundus (FDP) in all positions. The tendon of the small finger was the lowest among the group. Moment arms were slightly greater in neutral forearm position for each tendon. Significant difference was not observed with change in wrist position, for RUD. The tendon of FDS and FDP of the index and middle fingers showed small radial moment arms, while tendons for the ring and little fingers had larger ulnar moment arms, which were maximum in forearm supination (7.6 - 9.2 mm).

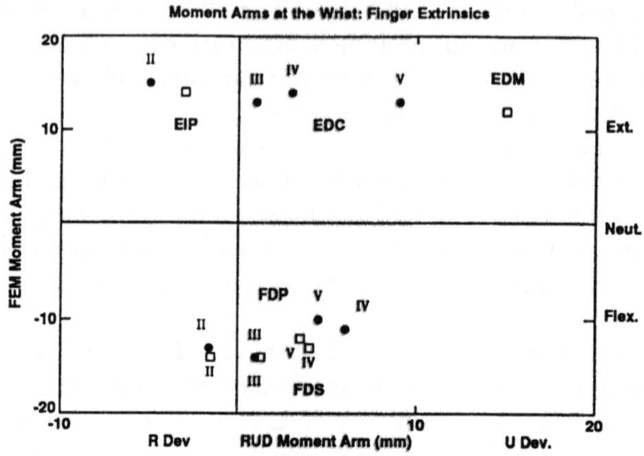

FIGURE 6. Cross-sectional view of thumb extrinsics with neutral wrist and forearm.

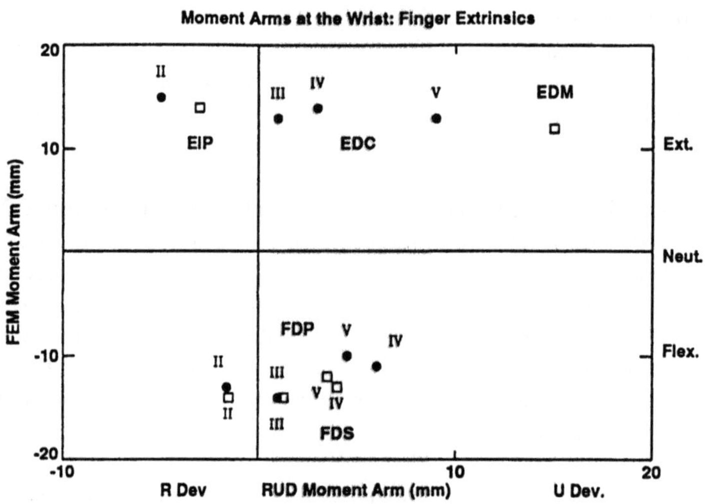

FIGURE 7. Cross-sectional view of finger extrinsics with neutral wrist and forearm (enlarged scale).

References

Agee J: (Personal communication)

Andrews JG, Youm Y: A biomechanical investigation of wrist kinematics. J Biomech 1977;12:83-93.

An KN, Takahashi K, Harrigan TP, Chao EYS: Determination of muscle orientations and moment arms. Journal of Biomechanical Engineering 1984;106:280-282.

An KN, Ueba Y, Chao EYS, Cooney WP, Linscheid RL: Tendon excursion and moment arm of index finger muscles. J Biomech 1983;16:419-425.

An KN, Hui FC, Morrey BF, Linscheid RL, Chao EYS: Muscle across the elbow joint: a biomechanical analysis. J Biomechanics 1981;14:659-669.

Armstrong TJ, Chaffin DB: An investigation of the relationship between displacements of the finger and wrist joint and the extrinsic finger flexor tendons. J Biomech 1977;11:119-128.

Berger RA, Blair WF, Andrews JG: Resultant forces and angles of twist about the wrist after ECRL to ECU tendon transfer. J Ortho Res 1988;6:443-451.

Brand PW: Clinical mechanics of the hand. St. Louis, Missouri, CV Mosby Company, 1985, pp.192-309.

Brand PW, Beach RB, Thompson DE: Relative tension and potential excursion of muscles in the forearm and hand. J Hand Surg 1981;6:209-219.

Green DP: Radial nerve palsy. In: Green DP, eds. Operative Hand Surgery 1. New York, Churchill Livingstone, 1988, pp.1479-1498.

Kaufman KR, An KN, Chao EYS: Incorporation of muscle architecture into muscle length-tension relationship. J Biomechanics 1989;22:943-948.

Landsmeer JMF: Studies in the Anatomy of Articulation 1. The equilibrium of the 'intercalated' bone. Acta Morph Netherlands Scand 1960;3:287-303.

Lieber RL, Fazeli BM, Botte MJ: Architecture of selected wrist flexor and extensor muscles. The J of Hand Surgery 1990;15A:244-250.

Ohnishi N, Ryu J, Colbaugh R, Rowen B: Tendon excursion and moment arm of wrist motors and extrinsic finger motors at the wrist. Presented in 45 the annual meeting of American Society for Surgery of the Hand, Toronto, Canada, September 24-27, 1990.

Otten, E: Morphometrics and force-length relations of skeletal muscles. International Series on Biomechanics (ISB) Biomechanics IX-A (Edited by Winter, et al.) pp.27-32. Human Kinematic Publishers, Champaign, Illinois, 1985.

Steno N: Elementorum myologiae specimen s. musculi descriptio geometrica p.108 in Opera Philosophico, Volume II (Edited by Vilhelm Maar) Copenhagen (1910). Quoted in Bastholm E. The History of Muscle Physiology, Copenhagen, Ejner Munksgaard (1950).

Volz RG, Lieb M, Benjamin J: Biomechanics of the Wrist. Clin Orthop 1980;149: 112-117.

Woittiez, RD, Huijing PA, Boom HBK, Rozendal RH: A three dimensional muscle model: a quantified relation between form and function of skeletal muscles. J Morph. 182:95-113.

Youm Y, Thambyrajah K, Flatt AE: Tendon excursion of wrist movers. J Hand Surg 1984;9A:202-209.

Epilogue

The last decade of the nineteenth century witnessed the dawn of a new age in our understanding of the musculoskeletal system due to the discovery and harnessing of x-rays. This was particularly true in the wrist, with Bryce's landmark treatise on carpal bone motion being published less than one year after the announcement of Roentgen's fantastic discovery. The concept of applying mechanical principles to the human musculoskeletal system gained credence in the 1960's. The next quantum leap of advancement in our understanding of the wrist began to evolve in the 1970's. It was during this period that applications of newly developed electronics, in the form of computers, transducers, imaging systems, etc., were applied to the wrist. Quantification of carpal kinematics and kinetics were now possible and a virtual explosion of information occurred. We are now at a point in time where we have the capacity and experience to sift through this new collection of data and reflect upon its validity and value to our understanding of carpal mechanics.

With the advancement of technology came a growing awareness of limitations imposed by the testing methods and attendant assumptions. We are now acutely aware of the pitfalls of our laboratory techniques so eloquently defined by Werner Heisenberg, the author of the uncertainty principle (the precursor to quantum mechanics), whereby the act of observing a system inherently changes that system. Questions have arisen regarding the direct applicability of in-vitro results to in-vivo situations. We still have not solved an apparently indeterminate solution of simultaneously loading tendons across the wrist in-vitro to uniquely position the wrist. When attempting to model pathologic conditions in cadaveric wrists in which the pathologic condition does not naturally exist, a common omission is validation of the model. There is a growing sense that we are about to break through to yet another level of laboratory sophistication, spawned in part by recent advances in imaging through magnetic resonance technology, ever increasing data manipulation capacity, and improvements in data collection devices, such as miniaturization of strain gauges and load transducers. A non-inclusive list of specific areas of future interest will be suggested below.

Anatomy

The great European anatomists described carpal anatomy during the last four centuries at a level that had been sufficient for previous levels of understanding of carpal mechanics and pathomechanics. However, it is becoming increasingly apparent that a more detailed understanding of the anatomy of the wrist be made to keep pace with questions arising from more detailed laboratory evaluations and a greater sense of clinical sophistication. Investigations currently underway are re-evaluating the gross and light-microscopic anatomy of the carpal ligaments. Future investigations should continue in this field, with emphasis on the three-dimensional infrastructure of the ligaments and how they change orientation and configuration as the wrist is moved. Studies are needed which explore the nature of strains in these ligaments as they relate to morphology. As in the anterior cruciate ligament of the knee, a complex network of nerves within the substance of some of the capsular carpal ligaments has been discovered. This raises questions regarding possible proprioceptive roles for these ligaments, which may be as important functionally as the mechanical properties of the ligaments. Additionally, investigations should continue to attempt to integrate anatomical features with mechanical models, such as joint surface geometry and soft tissue constraints.

Material Property Studies

As we gain more information about the anatomy of the ligaments of the wrist, investigations will need to be designed to relate that information to our mechanical understanding of the wrist. Very little work has been published regarding the material properties of the wrist ligaments. When designing such studies, significant effort should be directed at insuring that the experimental design is constructed to produce data which is "physiologically valid". Agreement should be sought regarding the parameters to be measured, the rates at which loads are delivered, and the direction displacement is programmed to occur. We should strive to develop methods of testing the carpal ligaments under conditions of cyclic loading, which will help test theories of progressive instability patterns. Additionally, cooperative efforts should be mounted with physicists and radiologists to develop the technology to detect changes in strain in connective tissue non-invasively.

Kinematics

Numerous methods of accurately quantifying carpal bone motion and overall wrist motion have been developed, but we are still plagued by the sense that we do not fully understand carpal bone kinematics. Perhaps this is due in part to the use of reference coordinate systems which have no direct anatomic relationship. Perhaps it is due to a lack of consistency in data collection and result presentation methods. Serious consideration must be given to several fundamental questions by all investigators interested in studying carpal bone kinematics. Decisions regarding the use of passive or simulated active tendon forces to produce wrist motion must be made and the method of simulating tendon forces should be standardized. Further standardization should be sought when defining the reference coordinate systems used to define the resultant motion. The use of unconstrained versus planar-constrained motion patterns will need to be addressed. For the consumer of information derived from these studies, standardization of display methods of results would be helpful. Although a significant amount of work remains in the area of defining normal carpal kinematics, it will be important to continue to add new variables to these studies such as disruption of ligaments, tendon transfers, arthroplasty techniques, etc. Care will need to be employed, however, to be certain that all attempts have been made to limit and understand as fully as possible the number of variables introduced and to make an effort to study clinically relevant situations. Exciting techniques using new imaging technology to generate intrinsic coordinate systems in-situ should be explored fully, with a goal of implementing these methods clinically to study in-vivo normal and pathokinematics. These imaging techniques have the added capacity of allowing visualization of soft tissues such as capsular ligaments as well. Mathematical modelling of changes in the configuration and orientation of ligaments relative to changes in carpal bone orientation will represent a major advance in understanding carpal bone kinematics.

Kinetics

Without doubt, the largest volume of new information on carpal mechanics has come from kinetic investigations. Recent descriptions of joint contact forces using pressure sensitive film have given us an exciting glimpse into changes which may occur within the radiocarpal joint with various simulations of clinical situations, such as scaphoid malrotation, variance of the ulna, etc. Investigators utilizing pressure sensitive film are becoming aware of limitations and precautions in its use, such a recognition of the time dependency of load application and limitations of resolution. Standardization of experimental protocols, slowly evolving to date, should be encouraged and published to allow new investigators to enter the field. Careful comparisons with newer methods of estimating joint contact forces through pressure transducers should be made. Additionally, studies are needed exploring joint contact forces in the mid-carpal and intercarpal articulations. Strain gauges have been implemented in studies of lunate and scaphoid deformation. Not only should further experimental work continue with these techniques to study other carpal bones and various pathologic variables, but efforts must be made to counter the "Heisenberg" effect by reducing the level of invasiveness currently necessary for the use of such devices. In part, this may be realized through expanded use of analytical modelling. A rigid body-spring model has been developed for two-dimensional force analysis of the wrist. Work should certainly continue in this area, as well as expansion to a three-dimensional model, possibly implementing aspects of finite element analysis. As always, investigators of carpal kinetics will keep a vigilant look-out for new techniques and instrumentation, developed for other purposes, which might serve future investigation of the wrist well.

Future Investigations

Laboratory investigation of the wrist remains in its infancy, but is poised for explosive growth. The number of new investigators continues to increase, new technology continues to surface allowing ever-more detailed analyses, and increasing demands for information from clinicians are all encouraging such investigations. Over the next decade, it will be incumbent upon all investigators to attempt to standardize experimental technique to allow comparison and validation of results between laboratories. As new devices and methods of investigation develop, an attempt to critically compare the results with established methodology should be carried out. A worrisome element of past biomechanics publications all too often has been conclusions drawn from studies which have tested an insufficient number of specimens or based on inadequate statistical analyses. A conscious effort must be made by all investigators to combat this, by collaborating with statisticians whenever feasible. Finally, the tremendous sense of cooperation between wrist investigators that has been enjoyed over the previous years must be nurtured, protected and encouraged. The value of continuing contributions to the advancement of our understanding of wrist mechanics by open discussion at conferences and personal communications cannot be overemphasized.

Richard A. Berger
Kai-Nan An
William P. Cooney, III
Mayo Clinic/Mayo Foundation
Rochester, Minnesota

MIX
Papier aus verantwortungsvollen Quellen
Paper from responsible sources
FSC® C105338

If you have any concerns about our products,
you can contact us on
ProductSafety@springernature.com

In case Publisher is established outside the EU,
the EU authorized representative is:
Springer Nature Customer Service Center GmbH
Europaplatz 3, 69115 Heidelberg, Germany

Printed by Libri Plureos GmbH
in Hamburg, Germany